The New York Times
Book of
LANGUAGE
AND
LINGUISTICS

Other books in the series:
The New York Times Book of Archeology
The New York Times Book of Birds
The New York Times Book of the Brain
The New York Times Book of Fish
The New York Times Book of Fossils and Evolution
The New York Times Book of Genetics
The New York Times Book of Insects
The New York Times Book of Mammals
The New York Times Book of Natural Disasters

The New York Times

Book of
LANGUAGE
AND
LINGUISTICS

EDITED BY

NICHOLAS WADE

THE LYONS PRESS
Guilford, Connecticut
An imprint of The Globe Pequot Press

Copyright © 1998, 2003 by The New York Times

Introduction and chapter prefaces © 1998, 2003 by Nicholas Wade

Originally published under the title *The Science Times Book of Language and Linguistics*

ALL RIGHTS RESERVED. No part of this book may be reproduced or transmitted in any form by any means, electronic or mechanical, including photocopying and recording, or by any information storage and retrieval system, except as may be expressly permitted in writing from the publisher. Requests for permission should be addressed to The Lyons Press, Attn: Rights and Permissions Department, P.O. Box 480, Guilford, CT 06437.

The Lyons Press is an imprint of The Globe Pequot Press

10 9 8 7 6 5 4 3 2 1

Printed in the United States of America

Designed by Joel Friedlander, Marin Bookworks

Library of Congress Cataloging-in-Publication Data

The New York times book of language and linguistics / edited by Nicholas Wade.
    p. cm.
    Rev. ed. of: The science times book of language and linguistics, 2000.
    Includes index.
    ISBN 1-58574-793-9 (pbk. : alk. paper)
    1. Language and languages. 2. Linguistics. I. Wade, Nicholas. II. New York times. III.
    Science times book of languages and linguistics.

P107.N48 2003
400—dc21

2003047410

Contents

Introduction 1

1   **The Tree of Language**   3

      Linguists Dig Deeper into Origins of Language 5
           JOHN NOBLE WILFORD

      Linguists Debating Deepest Roots of Language 10
           JOHN NOBLE WILFORD

      Gene Study Sees No Tie to Spread of Languages 16
           WILLIAM K. STEVENS

      When No One Read, Who Started to Write? 19
           JOHN NOBLE WILFORD

      New Family Tree Is Constructed for Indo-European 26
           GEORGE JOHNSON

      Scholars Debate Roots of Yiddish, Migration of Jews 31
           GEORGE JOHNSON

      Mummies, Textiles Offer Evidence of Europeans in Far East 37
           JOHN NOBLE WILFORD

      Luigi Luca Cavalli-Sforza; A Geneticist Maps Ancient Migrations 45
           LOUISE LEVATHES

2   **Language in Other Species**   51

      Look Who's Talking. Don't Bother Listening. 53
           JOHN NOBLE WILFORD

      Ancestral Humans Could Speak, Anthropologists' Finding Suggests 56
           JOHN NOBLE WILFORD

      Brain of Chimpanzee Sheds Light on Mystery of Language 60
           SANDRA BLAKESLEE

      Chimp Talk Debate: Is It Really Language? 64
           GEORGE JOHNSON

She Talks to Apes and, According to Her, They Talk Back 70
    CLAUDIA DREIFUS

Picking Up Mammals' Deep Notes 74
    JANE E. BRODY

3   The Acquisition of Language   81

In Brain's Early Growth, Timetable May Be Crucial 83
    SANDRA BLAKESLEE

Linguists Debate Study Classifying Language as Innate Human Skill 90
    GINA KOLATA

Babies Learn Sounds of Language by Six Months 95
    SANDRA BLAKESLEE

Study Finds Baby Talk Means More Than a Coo 98
    SANDRA BLAKESLEE

Studies Show Talking with Infants Shapes Basis of Ability to Think 101
    SANDRA BLAKESLEE

Deaf Babies Use Their Hands to Babble, Researcher Finds 107
    NATALIE ANGIER

Removing Half of Brain Improves Young Epileptics' Lives 110
    ABIGAIL ZUGER

Test-Tube Moms 113
    STEPHEN S. HALL

When an Adult Adds a Language, It's One Brain, Two Systems 119
    SANDRA BLAKESLEE

Old Brains Can Learn New Language Tricks 122
    SANDRA BLAKESLEE

Scientists Track the Process of Reading Through the Brain 126
    GINA KOLATA

4   Language and the Brain   131

Brain Yields New Clues on Its Organization for Language 134
    SANDRA BLAKESLEE

The Mystery of Music: How It Works in the Brain 141
    SANDRA BLAKESLEE

Odd Disorder of Brain May Offer New Clues 149
    SANDRA BLAKESLEE

Contents

Brain May Have Separate Units to Digest Writing and Speech 155
    SANDRA BLAKESLEE

Subtle But Intriguing Differences Found in the Brain Anatomy of Men and Women .. 158
    DANIEL GOLEMAN

Men and Women Use Brain Differently, Study Discovers 163
    GINA KOLATA

5   Language and Society   167

Laughs: Rhythmic Bursts of Social Glue 169
    NATALIE ANGIER

Time Almost Buried Ancient Maya Secrets 176
    JOHN NOBLE WILFORD

Language of Early Americans Is Deciphered 178
    JOHN NOBLE WILFORD

In a Publishing Coup, Books in "Unwritten" Languages 184
    JOHN NOBLE WILFORD

Indians Striving to Save Their Languages 189
    JAMES BROOKE

6   The Latest from the Field   195

What We All Spoke When the World Was Young 198
    NICHOLAS WADE

Expert Says He Discerns "Hard-Wired" Grammar Rules 205
    BRENDA FOWLER

Researchers Say Gene Is Linked to Language 209
    NICHOLAS WADE

Language Gene is Traced to Emergence of Humans 213
    NICHOLAS WADE

Introduction

"Whereof one cannot speak," said the philosopher Wittgenstein, "thereof one must remain silent." Language is not just the spoken and written word, it is also the currency of conscious thought. What one cannot put into words is hard even to think about.

Humans do not have a monopoly of thought, but language seems to be the one biological ability that we alone possess, at least among living species.

Our distant relatives, the chimpanzees, are highly intelligent and in captivity can learn a large repertoire of signs. But they lack the vocal apparatus necessary for language and the neural centers to string the signs together in anything resembling syntax.

Extinct species of hominid, like our probable ancestors *Homo habilis* and *Homo ergaster,* or our cousins the Neanderthals, may have spoken, but there is no sure evidence that they did.

About 50,000 years ago, after millennia of steady change, the archeological record reveals a spectacular flowering of the human behavioral repertoire, including the first art objects and an accelerating pace of technological innovation. Some archeologists, like Richard G. Klein of Stanford University, believe that some genetically based neural change preceded the emergence of these behaviorally modern humans.

Given the pace of genetic change that underlay hominid evolution, including a steady increase in the size of the brain, the possibility of a genetic change that influenced behavior is not implausible. But what change could have influenced behavior so profoundly? Dr. Klein believes it was one that promoted the development of fully modern speech.

If the language dates back 50,000 years ago to the emergence of behaviorally modern humans, and if the original population of modern humans was small and cohesive enough, then a single language may have been spoken by everyone.

The myth of the Tower of Babel can be seen as a metaphor for the spread of humans throughout the globe, an event that seems to have begun about 40,000 years ago when a small breakaway group emigrated from the African continent. As different populations lost touch with one another, each band's language would presumably have been free to develop without constraint. Given the inherent changeability of language, the single tongue soon became thousands, and the gift of language that had allowed all humans to share their thoughts now reinforced the divisions.

Some brave linguists hope to work back from present-day languages and reconstruct the ancient mother tongue. But most believe language changes too quickly for its roots to be traced more than a few thousand years. There is general agreement that English and most other European languages are derived from an ancestral tongue that scholars call proto-Indo-European. But historical linguists differ as to whether the tree of language can with confidence be traced any farther back.

Since the gift of language is uniquely human, its acquisition and the vocal and neural structures that support it are of particular interest. In recent years, developmental psychologists and neurophysiologists have made important strides in understanding how the brain is organized for language, but much remains to be learned.

No language lives forever; they become extinct, when those that speak them are conquered or perish, or fade into obsolescence as each generation chooses to speak in a slightly different idiom from its parents. From written inscriptions, archeologists can occasionally rescue a lost language from oblivion. But languages are dying now in greater numbers than ever before as primitive peoples become acculturated into the civilizations that encroach on their distant valleys or forests.

The following articles, which report recent finding on the various aspects of language, appeared first in the science section of the *New York Times*. My colleagues on the science section and I would like to thank Lilly Golden of the Lyons Press for proposing the idea of this book, giving the ephemeral words of daily journalism a new life between hard covers.

— NICHOLAS WADE, Spring 1999

1

THE TREE OF LANGUAGE

Historical linguistics is the effort to reconstruct the ancient roots of today's languages and, in its most ambitious form, to figure out the hypothetical mother tongue from which all languages are descended.

Genetic and geographic studies pioneered by Luca Cavalli-Sforza show that a small group split away from the ancestral human population of Africa and spread to Asia. The Asian population turned west to colonize Europe and continued its eastern migration across the Bering Strait to populate the Americas.

As these early human populations dispersed, to each community language development would have followed an independent course, leading to the babel of tongues seen around the world today.

The rapidity of linguistic change is evident from the language family that has been most closely studied, that of Indo-European. Swedish and Iranian do not have many obvious similarities, but linguists can trace the development of each to a single mother tongue. The language, known as proto-Indo-European, has long since faded into oblivion, but many of its words can be inferred.

If most of the languages of Europe can be traced to a single tongue, spoken probably some 10,000 years ago, why cannot the thread be followed farther back in time? Many linguists believe language changes so rapidly that the relationships between kindred tongues inevitably disappear, and that the original language of humankind, if it existed, is irretrievably lost.

But a minority of linguists, led by Joseph Greenberg and Merritt Ruhlen, believe that the major language families can plausibly be united. "Nostratic" is the name of a hypothetical language ancestral to many European and Asian languages and supposed to have been spoken some 25,000 years ago.

Linguists Dig Deeper into Origins of Language

PALEONTOLOGISTS PONDER THEIR FOSSILS, archeologists turn over ancient stones and now scholars of linguistics are joining the search for human origins with a systematic analysis of the roots of the world's known languages. They are seeking ultimately to reconstruct the primordial language, the mother tongue of all humans.

Not that anyone hopes to find some Rosetta stone for the earliest ancestral language. The written word goes back only 6,000 years. Yet humans may have developed rudimentary spoken language at least 50,000 years ago, although from the evidence of fossil jaws, they probably could not have made the sound of any vowels other than a long "a."

And not that anyone expects to learn that 30,000 years ago Cro-Magnon artists discussed their cave paintings in words bearing any resemblance to modern languages. Over time, even a single millennium, languages change beyond easy recognition; the old English of *Beowulf* in the eighth century is practically unintelligible, and 14th-century Chaucer is no snap.

However, a few linguistic experts now firmly believe that, by analyzing the evolution and approximate sound and meaning of certain words, they have reconstructed the basic vocabulary of a remote ancestral language that modern man has never heard or seen. They call it Nostratic, from the Latin *noster*, for "our." Nostratic was spoken in the Middle East sometime between 20,000 and 12,000 years ago, the researchers say, and from it evolved all the European languages as well as many African and Asian ones.

Other linguists, employing the same meticulous analytical techniques, report that they have reconstructed the root languages of the first settlers of the Americas, the ancestors of the Eskimos and Indians. The knowledge could point to the origin in Asia of the first Americans and their migration patterns.

Emboldened by these reported successes, Vitaly V. Shevoroshkin, a linguist at the University of Michigan, is striving to determine some of the words of an even earlier protolanguage, the common root of Nostratic and other seemingly unrelated linguistic branches on the family tree of prehistoric language. He believes this ancestral language, spoken 25,000 years ago, could eventually lead him to discover some of the elements of what may have been the primordial language.

"Ultimately, all languages, with perhaps some little exceptions, are related," Dr. Shevoroshkin said, stating the guiding premise of those scholars who are making a scientific study of the distant relationship, of languages.

The search for such relationships is potentially valuable to the larger study of human origins by anthropologists, paleontologists and archeologists. Finding linguistic similarities back through time could buttress other research on the human migrations from Africa to the Middle East and then throughout the world. Ancestral vocabularies could provide clues to the tools early peoples used, the animals they domesticated and their means of survival.

Winfred P. Lehmann, a retired professor of linguistics and Germanic languages at the University of Texas at Austin, said, "We can learn more about prehistory through language, possibly where civilization actually developed. Words give us a notion of what people were talking about, and thus something about their culture."

Like many linguistic scholars, Dr. Lehmann is skeptical of some of the conclusions reached by the "Nostratic School," as the seekers of the earliest protolanguages are called. This research was pioneered in the 1960s in the Soviet Union, where most of the work is still done. Dr. Shevoroshkin, the leading exponent in this country, emigrated here 13 years ago.

The Nostratic scholars are "very ingenious and very capable," Dr. Lehmann said. "You have to respect the idea that all the languages were related twenty-five thousand years ago. But the grounds for their conclusions are not always convincing. If you go back a few thousand years, comparing words and languages, pretty soon you have so few related words left that you have nothing to compare."

For this reason, linguists have generally despaired of ever tracing the lineage of languages back beyond about 5000 B.C., and are dubious of the new attempts to do so.

Relationships of modern languages are usually apparent. Similarities among Italian, Spanish, Portuguese, French and Romanian betray their Latin origin. English is a member of the Germanic family, and even the timing and circumstance of its origin are well documented; Anglo-Saxons speaking an Old German invaded the British Isles in the fifth century A.D. and, living in isolation, evolved their separate but related language.

Scholars in the 19th century made the first leap back in hypothetical ancestral languages. By comparing archaic words of modern languages and analyzing internal vowel changes and common word endings, they established the probable common root of Latin, Greek, Sanskrit, Germanic, Celtic, Balto-Slavic and Indo-Iranian. Called Indo-European, its daughter languages constitute today's most widely spoken linguistic family. One of the 6,000 words in the Indo-European dictionary may be transliterated approximately as *peter*, for "father"—compared with *pitar* in Sanskrit and *pater* in Latin.

For a few thousand years after 5000 B.C., Indo-European was probably spoken in parts of the Middle East and around the Caspian and Black seas. Sanskrit evolved from it about 1500 B.C. or earlier; Greek, about 1450 B.C.

Contemporary with Indo-European, it was later determined, were nine other root protolanguages, including Afro-Asiatic (the source of Arabic and Hebrew, among others), Uralic (Finnish and Hungarian) and Altaic (Mongolian, Japanese and Korean).

Whether it can be proved that Indo-European and these other protolanguages were related, through a common tongue back in time, is a matter of dispute among linguists, except for disciples of the Nostratic School.

In 1963, two Soviet linguists, Vladislav M. Illich-Svitych and Aaron Dolgopolsky, independently arrived at some words that they said were probably in the vocabulary of the common ancestral language about 12,000 years ago. The Nostratic dictionary has now been expanded to more than 500 words.

Dr. Shevoroshkin, who was a colleague of Dr. Dolgopolsky before they both left the Soviet Union, said the reconstruction of such an ancestral language began with the attempt to find in each descendant language words and phonetic patterns that were the most archaic, and hence probably from an earlier language. An important step in this analysis was determining the most stable words through time, which Dr. Dolgopolsky did after examining 140 languages of Europe and Asia. "Using statistical methods, Dolgopolsky identified and inventoried the fifteen stablest meanings in these languages," Dr. Shevoroshkin explained. "These were the words that could then be studied for evidence of common ancestry once similarities based on pure coincidence and borrowings had been ruled out."

These words represent meanings that are seldom or never replaced by other words with the same meaning in any given language. They were mostly nouns, including words for many body parts.

The stablest 15 meanings, according to their list, were words for the following objects or concepts: I/me; two/pair; thou/thee/you; who/what; tongue; name; eye; heart; tooth; no/not; fingernail/toenail; louse; tear (as in weeping); water; and dead.

By studying words with these meanings in many languages, and then expanding the list, the scholars began to see similarities in the use of vowels and consonants and general patterns for shifts in the typical sounds that make up words.

In this way, Soviet scholars established that Nostratic for "I" was *mi*, and for "me" was *mi-nV*—the upper-case "V" denoting a vowel whose sound is undetermined. In Indo-European, it was *me* and *mene*, and similar words have survived in modern languages.

The Nostratic *kuni*, for "wife" or "woman," becomes in Altaic *kuni*, in Afro-Asiatic *KwVn* (the upper-case "K" symbolizes a glottal sound) and in Indo-European *gwen*. The Indo-European word survived into Middle English and is the root of the English "queen."

Dr. Shevoroshkin complains that out of ignorance and skepticism scholars in the United States are discouraged from pursuing Nostratic techniques for reconstructing protolanguages. His applications for research grants have been repeatedly rejected. When he taught at Yale University, before joining the Michigan faculty, he said he was told "not to discuss Nostratic theory in my classes."

Soviet linguists continue the research, however, and have reconstructed another protolanguage that was probably contemporary to Nostratic. It is called "Dene-Caucasian," the mother tongue of Chinese, among other Asian languages. One of its offshoots is believed to be the Na-Dene family of languages, whose speakers were among the first migrants to America.

Joseph Greenberg, a retired professor of linguistics at Stanford University, has applied similar analytical techniques to try to demonstrate that all the languages of the native Americans stemmed from three protolanguages. Besides Na-Dene, he says, there was Eskimo-Aleut, which derived from Nostratic and is therefore closer to the tongue of America's eventual conquerors, and an independent grouping that Dr. Greenberg calls "Amerind."

According to Dr. Greenberg and his colleague, Merritt Ruhlen, the fact that the Amerind family of languages is more widely scattered in both North and South America indicates that these were probably the first migrants to the New World.

In a recent article in *Natural History* magazine, Dr. Ruhlen, the author of *A Guide to the World's Languages*, published this year by Stanford University Press, wrote: "There is strong evidence for three migrations in that each of the three New World families appears to be more closely related to language families in the Old World than to either of the other two New World families."

Dr. Ruhlen, in an interview, conceded that reconstructed protolanguages were "educated guesses based on a range of meanings and a range of sounds," but he said that many critics are often unfamiliar with the Nostratic methods, in part because most of the research has been published in Russian. In 1986, Dr. Shevoroshkin and Thomas L. Markey, another Michigan linguist, published a book of translated essays on Nostratic theory.

Visitors to Dr. Shevoroshkin's office in Ann Arbor are handed a poem written in Nostratic that conveys the sense of promise felt by those who search for protolanguages, as well as recognition of the controversy surrounding their efforts. Roughly translated, the poem reads:

> Language—ford through river of time,
> It leads us to the dwelling of dead,
> But those cannot arrive there,
> Who are afraid of deep water.

—JOHN NOBLE WILFORD, November 1987

Linguists Debating Deepest Roots of Language

IN THEIR ARCHEOLOGICAL DIGS THROUGH the strata of human language, linguists have long been fascinated by the seeming similarities between the English words "fist," "finger" and "five." The motif is repeated by the Dutch, who say "vuist," "vinger" and "vijf," and the Germans, who say "faust," "finger" and "funf." Traces of the pattern can even be found as far away as the Slavic languages like Russian.

Conceivably, sometime in the distant past, before these languages split from the mother tongue, there was a close connection among the words for a hand and its fingers and the number five. But did the mathematical abstraction come from the word for fist, or, as some linguists have proposed, was it the other way around? The answer could provide a window into the development of the ancient mind.

In a paper being prepared for publication in a book, Dr. Alexis Manaster Ramer, a linguist at Wayne State University in Detroit, argues that the mystery may now be solved: "fist" came before "five." But more important than his conclusion is the method by which it was derived.

It is widely accepted that English, Dutch, German and Russian are each branches of the vast Indo-European language family, which includes the Germanic, Slavic, Romance, Celtic, Baltic, Indo-Iranian and other languages—all descendants of more ancient languages like Greek, Latin and Sanskrit. Digging down another level, linguists have reconstructed the even earlier tongue from which all these languages are descended. They call it "proto-Indo-European," or PIE for short.

But in a move sure to be hotly disputed by mainstream linguists, Dr. Manaster Ramer contends that to find the root of the fist-five connection one must look beyond the Indo-European family and examine two separate language groups: Uralic, which includes Finnish, Estonian and Hungarian, and Altaic, said to include Turkish and Mongolian languages . All three families, he contends, contain echoes of a lost ancient language called "Nostratic."

If Dr. Manaster Ramer is right, his discovery will provide ammunition for a small group of linguists who make the controversial claim that Indo-European, Uralic, Altaic and other language families like Afro-Asiatic, which includes Arabic and Hebrew, the Kartvelian languages of the South Caucasus and the Dravidian languages concentrated in southern India, all are descendants of Nostratic, which was spoken more than 12,000 years ago.

Most language experts remain highly skeptical of the Nostratic hypothesis, which enjoyed so much publicity in the late 1980s and early 1990s that it is sometimes described as the linguists' version of cold fusion. "It would be terrific if it's true, but we don't want to jump to conclusions," said Dr. Brian Joseph, a linguist at Ohio State University in Columbus. Dr. Joseph and Dr. Joe Salmons of Purdue University in West Lafayette, Indiana, are editing the book *Nostratic: Evidence and Status* (John Benjamins), in which the analysis of the five-fist connection will appear.

But Dr. Joseph believes that while the Nostratic debate remains as heated as ever, it has reached a higher level of sophistication, with both sides offering more precise arguments and careful scholarship. "Mainstream linguists who in the past had dismissed Nostratic are now willing to examine it on an objective and scientific basis," he said. While he and Dr. Salmons both count themselves as skeptics, they hope their book will be a milestone in linguistic scholarship. "Even if the more mainstream linguists decide to reject Nostratic," Dr. Joseph said, "at least the evidence will be laid out in a fair and balanced way."

It is not that most linguists find implausible the idea that all languages may ultimately have derived from an ancient ur-language spoken millennia ago. After all, analysis of mitochondrial DNA from the cells of various ethnic groups strongly supports the notion that all humans come from the same genetic stock. If this small group of original humans spoke a single language, then all present-day languages are descended from it. The hypothetical Nostratic is not the ur-language but might be one of its major branches. However, critics of the Nostratic hypothesis have long argued that it is unprovable—any similarities between languages as distant as the Altaic and Indo-European would have been washed out long ago. They dismiss the parallels unearthed by the Nostraticists as coincidences.

As recently as the early 1990s, most evidence for an ancient language relied on work done in the 1960s by Soviet scholars, who coined the word "Nostratic" to mean "our language." But now a second wave of research is revitalizing the field. Veterans of the Nostratic program like Dr. Vitaly V. Shevoroshkin of the University of Michigan in Ann Arbor and Dr. Aaron Dogopolsky of the University of Haifa in Israel continue to come up with new evidence, as do younger scholars like Dr. Manaster Ramer.

In a book published in 1994, *The Nostratic Macrofamily: A Study in Distant Linguistic Relationship* (Mouton de Gruyter), two independent scholars, Allan Bomhard and John Kerns, compiled some 600 Nostratic roots with counterparts (what the linguists call "cognates") in languages said to be descended from Nostratic. On another front, Dr. Joseph Greenberg, a retired Stanford University linguist, is in the midst of a two-volume study of his own version of the Nostratic hypothesis entitled *Indo-European and Its Closest Relatives: The Eurasiatic Language Family.* Dr. Greenberg's Eurasiatic overlaps with Nostratic but also includes other languages like Japanese and Eskimo-Aleut.

In an unpublished manuscript of yet another forthcoming book, *Indo-European and the Nostratic Hypothesis,* Mr. Bomhard concludes that the evidence for the common ancestral language is "massive and persuasive." "As the twentieth century draws to a close, it is simply no longer reasonable to hold to the view that Indo-European is a language isolate," he writes. "Indo-European has relatives and these must now be taken into consideration."

One of the most vehement critics of Nostratic, Dr. Donald Ringe, a linguist at the University of Pennsylvania, recently surprised himself by finding statistical evidence that resemblances between Uralic and Indo-European may indeed be due to more than chance. Dr. Ringe expected his analysis, which will also be published in the book edited by Dr. Joseph and Dr. Salmon, to undermine the Nostratic hypothesis.

But Dr. Ringe is quick to point out that a connection between Indo-European and the Uralic languages like Hungarian and Finnish is the least controversial claim of the Nostraticists. He remains as dubious as ever that statistically significant connections can be found between Indo-European and more distant languages.

In a paper called "'Nostratic' and the Factor of Chance," published in the journal *Diachronica,* Dr. Ringe examined a list of 205 cognates that the Russian linguist Vladislav Illich-Svitych found among six language families commonly said to have descended from Nostratic. He concluded that the similarities are indistinguishable from those that would have arisen by chance. As a test of his analytical technique, Dr. Ringe applied the same method to two Indo-European languages that are known to be related, and found that the similarities there are indeed statistically significant.

"It is time to tighten up standards of evidence in historical linguistics," he concluded in his paper. "If we enforce rigor, the truth will enforce itself."

But some linguists believe Dr. Ringe is misinterpreting his own statistics. Dr. Manaster Ramer argues that Dr. Ringe, who has accused the Nostraticists of "innumeracy," is himself engaging in "pseudomathematics."

"To use mathematics in any science, including linguistics, you have to understand the meaning of the mathematics and not just learn to manipulate formulas," he said. Dr. Manaster Ramer believes that the probability distribution that Dr . Ringe found for Nostratic is exactly what would be expected in languages that split apart long ago and developed independently. The true test of whether languages are related is not statistical comparisons, he insists, but the tools of historical linguistic analysis. If one can find answers in Uralic and Altaic to puzzles in Indo-European, like the five-fist connection, he says, that strengthens the argument for an ancestral Nostratic tongue.

Historical linguists start with two languages they suspect are related, then search for potential cognates—words like the Italian *luce* ("light") and *pace* ("peace"), which appear in Spanish as *luz* and *paz*. Then, by deriving rules for how sounds mutate over time, they try to reconstruct the ancient roots: Latin *lux* and *pax*. German *vater* and English *father* can be traced to Latin *pater.*

In actual practice, the correspondences between related words are usually far more convoluted and opaque to superficial examination. English and Armenian both are believed to descend from proto-Indo-European. But it takes a great deal of linguistic manipulation to show how the Armenian word for "two," *erku,* is related to its English counterpart. To add to the confusion, words that seem similar can turn out to be unrelated. Linguists consider it coincidental that the German word for "awl" happens to be *ahle,* or that the Aztec word for "well" is *huel.* For that matter, the English word *ear,* referring to the fleshy flaps on either side of the head, has been found to be historically unrelated to an "ear" of corn.

There are other mirages that can create the illusion of a deep historical wellspring. Baby words like "papa" and "mama" are common across languages probably because the labial consonants—those made with the lips—are among the first that children learn. Onomatopoeic words like "clash" or "meow" also tend to turn up independently in unrelated languages . And of course languages borrow words from one another all the time. A Japanese office worker can log off her *konpyuutaa* and head for *Makudonarudo* to grab a *hanbaagaa* and a steaming cup of *hotto kohii* for lunch.

To avoid being misled by such specious similarities, linguists try to concentrate on basic words—numbers, parts of the body—likely to have been embedded in a language from the start. As reconstructed by linguistic archeologists, the ancient Indo-European word for "five" was *penkwe,* which became *pende* in Greek, *quinque* in Latin and *panca* in Sanskrit. One can immediately see surface similarities between *penkwe* and the Indo-European roots for "fist," *pnkwstis,* and "finger" *penkweros.* But though the resonances ring, the source of the connection has remained obscure.

Finding few clues within Indo-European itself, Dr. Manaster Ramer looked farther afield. Linguists examining Finnish, Hungarian and Estonian had reconstructed an ancient Uralic root, *peyngo,* meaning "fist" or "palm of the hand." And from Turkish, Mongolian and related languages, linguists had reconstructed the corresponding word in Altaic: "p'aynga." (The accent is a sign that there were two different p sounds in the language.)

Working backward from Uralic and Altaic, Dr. Manaster Ramer reconstructed a hypothetical Nostratic antecedent, *payngo.* Then, using what he believed to be the rules by which Nostratic mutated into proto-Indo-European, he showed how the Nostratic word for "fist" could have spawned the Indo-European word for "five."

In another attempt to show that the Indo-European languages descended from Nostratic, Dr. Manaster Ramer analyzed the word "stink," which came into English from the hypothetical Germanic root *stinkwan*. Linguists find this word interesting because it appears to have no counterparts in other Indo-European languages. Dr. Manaster Ramer argues that it could have derived from a hypothetical Nostratic word, *stunga*.

In his own work, Mr. Bomhard points to evidence that the first-person pronoun "me" and variations like *mi, ma, mo* and *mea* appear in PIE and in the reconstructed protolanguages Kartvelian, Afro-Asiatic, Uralic, Altaic and the extinct language Sumerian. Mr. Bomhard believes that ancestral Indo-Europeans said *bor* for "to bore" or "to pierce"; the Afro-Asiatics said *bar,* the Altaics said *bur,* the Sumerians *bur,* and the Dravidians *pur,* while the Uralics said *pura* for "borer" or "auger." And while Indo-Europeans said *pes* or *pos* for "penis," speakers of Altaic said *pusu* for "to squirt out" or "to pour" and the Sumerians said *pes* not only for sperm and semen but also for descendant, offspring and son.

Most linguists are leery of reading too much significance into reconstructions that are based on reconstructions. Are the Nostraticists excavating into the past or building a house of cards?

"The bottom line is that the evidence isn't good enough," Dr. Ringe said. "In particular, neither Manaster Ramer nor anyone else has demonstrated that the similarities they've found between the various recognized language families are due to anything other than chance."

With such different ideas about how Nostratic scholarship should proceed, it is unlikely that either Dr. Ringe or Dr. Manaster Ramer will come around to the other's point of view. In the meantime, linguists watching from the sidelines say there is a huge amount of work to be done before Nostratic can confidently be verified or rejected.

"I think there is a new appreciation of the level of sophistication one needs to approach the problem," said Dr. Brent Vine, a Princeton University classicist. "So much is now known about all the different language families involved that no one person can seriously claim to have the kind of control needed for Nostratic research. What is really needed is a team effort."

—JOHN NOBLE WILFORD, November 1987

Gene Study Sees No Tie to Spread of Languages

A NEW ANALYSIS OF GENETIC patterns among modern Europeans has failed to support, but does not disprove, either of two competing theories on how Indo-European languages like English, French and German spread across Europe thousands of years ago, researchers at the State University of New York at Stony Brook say.

The conventional theory holds that the Indo-European languages were imposed on the early inhabitants of Europe by the conquerors who swept out of the steppes of what is now the southern Ukraine beginning about 6,500 years ago. A competing view says that the languages were spread by early farmers as they moved, century by century, in search of new lands, starting in the Middle East about 9,000 years ago.

Last year, a team headed by Dr. Robert R. Sokal lent support to the agricultural theory by demonstrating that genetic patterns are correlated with the spread of agriculture. If no correlation had been found, Dr. Sokal said, the agricultural theory would have collapsed because there would have been no mechanism for the spread of the languages.

But the existence of a mechanism does not prove the theory, Dr. Sokal said. And now, carrying the analysis a step farther, he and two colleagues at Stony Brook, Dr. Neal L. Oden and Barbara A. Thomson, have attempted to establish a similar correlation between genetics and the spread of language itself, as postulated by the two theories. There is none, they report in *The Proceedings of the National Academy of Sciences*.

"We have not disproved them," Dr. Sokal said of the competing theories, "but our genetic evidence does not support them. This argument has to shift to other ground."

In its first study, the Sokal team analyzed proteins from modern people at more than 3,300 sites across Europe. The researchers found that certain genes became progressively less common in going from southern Turkey, near the area where agriculture originated, toward northern Europe.

That genetic gradient was presumed to have come about because the genes of the original population were diluted as the group moved across Europe and intermarried with the hunter-gatherer populations along its path. The migration routes suggested by the genetic gradient showed a strong statistical correlation with the spread of agriculture, as known from the archeological record.

The correlation suggested, in turn, that genetic and cultural change moved in tandem from Turkey through the Balkans, proceeding northwestward with the spread of the new technique of agriculture, which enabled populations to grow and prompted them to search constantly for more land. The theory that agriculture was diffused by people was recently adopted by Dr. Colin Renfrew, an archeologist at Cambridge University in England, as support for his novel theory that the Indo-European languages were spread as a cultural adjunct to the spread of agriculture.

That theory ran directly counter to the one advocated for 20 years by, among others, Dr. Marija Gimbutas of the University of California at Los Angeles. It holds that the homeland of the Indo-European peoples and languages lay in the Pontic Steppes north of the Black Sea. These warlike invaders made their way into Europe in three waves starting around 6,500 years ago, according to this theory, imposing their language on the conquered as they went.

Many languages—perhaps Etruscan is the best known—predated the appearance of the Indo-European tongues. Today only one remains: the Basque language found in parts of Spain and France.

In the most recent phase of its analysis, the Sokal team attempted to match the genetic gradient to language patterns in Europe. The researchers say they have found a correlation between genetics and language, but no statistically significant evidence that the matchup is explained either by the language-follows-agriculture theory or the conquerors-out-of-Ukraine theory.

While Dr. Sokal has not proposed any alternative hypothesis, he said the results of the new analysis were consistent with some kind of branching theory. That is, populations may have split into different streams and moved to different regions, and their language likewise evolved in different directions.

"We find indirect evidence that there was a branching process going on," Dr. Sokal said. "Where this went on during the evolution of the Indo-European populations, we have no way of knowing. It could have happened outside Europe, with the descendants migrating into Europe, or it could have happened within Europe."

—WILLIAM K. STEVENS, August 1992

When No One Read, Who Started to Write?

THE SUMERIANS HAD A STORY to explain their invention of writing more than 5,000 years ago. It seems a messenger of the king of Uruk arrived at the court of a distant ruler so exhausted from the journey that he was unable to deliver the oral message. So the king, being clever, came up with a solution. He patted some clay and set down the words of his next messages on a tablet.

A Sumerian epic celebrates the achievement:

Before that time writing on clay had not yet existed,
But now, as the sun rose, so it was!
The king of Kullaba Uruk had set words on a tablet, so it was!

A charming just-so, or so-it-was, story, its retelling at a symposium on the origins or writing, held at the University of Pennsylvania, both amused and frustrated scholars. It reminded them that they could expect little help—only a myth—from the Sumerians themselves, presumably the first writing people, in understanding how and why the invention responsible for the great divide in human culture between prehistory and history had come about.

The archeologists, historians and other scholars at the meeting smiled at the absurdity of a king's writing a letter that its recipient could not read. They also doubted that the earliest writing was a direct rendering of speech. Writing more than likely began as a separate and distinct symbolic system of communication, like painting, sculpture and oral storytelling, and only later merged with spoken language.

Yet in the story, the Sumerians, who lived in Mesopotamia, the lower valley of the Tigris and Euphrates rivers in what is now southern Iraq, seemed to understand writing's transforming function. As Dr. Holly Pittman, director of the university's Center for Ancient Studies and organizer of the symposium, observed, writing "arose out of the need to store information and transmit information outside of human memory and over time and over space."

In exchanging interpretations and new information, the scholars acknowledged that they still had no fully satisfying answers to the most important questions of exactly how and why writing was developed. Many of them favored a broad explanation of writing's origins in the visual arts, pictograms of things being transformed into increasingly abstract symbols for things, names and eventually words in speech. Their views clashed with a widely held theory among archeologists that writing grew out of the pieces of clay in assorted sizes and shapes that Sumerian accountants had used as tokens to keep track of livestock and stores of grain.

The scholars at the meeting also conceded that they had no definitive answer to the question of whether writing was invented only once and spread elsewhere or arose independently several times in several places, like Egypt, the Indus Valley, China and among the Olmecs and Maya of Mexico and Central America. But they criticized findings suggesting that writing might have developed earlier in Egypt than in Mesopotamia.

In December, Dr. Gunter Dreyer, director of the German Archeological Institute in Egypt, announced new radiocarbon dates for tombs at Abydos, on the Nile about 250 miles south of Cairo. The dates indicated that some hieroglyphic inscriptions on pots, bone and ivory in the tombs were made at least as early as 3200 B.C., possibly 3400. It was now an "open question," Dr. Dreyer said, whether writing appeared first in Egypt or Mesopotamia.

At the symposium, Dr. John Baines, an Oxford University Egyptologist who had just visited Dr. Dreyer, expressed skepticism in polite terms. "I'm suspicious of the dates," he said in an interview. "I think he's being very bold in his readings of these things."

The preponderance of archeological evidence has shown that the urbanizing Sumerians were the first to develop writing, in 3200 or 3300 B.C. These are the dates for many clay tablets with a protocuneiform script found at the site of the ancient city of Uruk. The tablets bore pictorial symbols for the names of people, places and things for governing and commerce. The Sumerian script gradually evolved from the pictorial to the abstract, but it was probably at least five centuries before the writing came to represent recorded spoken language.

Egyptian hieroglyphics are so different from Sumerian cuneiform, Dr. Baines said, that they were probably invented independently not long after Sumerian writing. If anything, the Egyptians may have gotten the idea of writing from the Sumerians, with whom they had contacts in Syria, but nothing more.

In any event, the writing idea became more widespread at the beginning of the third millennium B.C. The Elamites of southern Iran developed a protowriting system then, perhaps influenced by the protocuneiform of their Sumerian neighbors, and before the millennium was out, writing appeared in the Indus River Valley of what is now Pakistan and western India, then in Syria and Crete and parts of Turkey. Writing in China dates back to the Shang period toward the end of the second millennium B.C., and it dates to the first millennium B.C. in Mesoamerica.

Archeologists have thought that the undeciphered Indus script, which seemed to appear first around 2500 B.C., may have been inspired in part from trade contacts with Mesopotamia. But new excavations in the ruins of the ancient city of Harappa suggest an earlier and presumably independent origin of Indus writing.

In a report from the field, distributed on the Internet, Dr. Jonathan Mark Kenoyer of the University of Wisconsin and Dr. Richard H. Meadow of Harvard University showed pictures of marks incised on potshards that they interpreted as evidence for the use of writing signs by Indus people as early as 3300 B.C. If these are indeed protowriting examples, the discovery indicates an independent origin of Indus writing contemporary with the Sumerian and Egyptian inventions.

Dr. Meadow, using E-mail, the electronic age's version of the king of Uruk's clay tablet, confirmed that the inscribed marks were "similar in some respects to those later used in the Indus script." The current excavations, he added, were uncovering "very significant findings at Harappa with respect to the Indus script."

At the symposium, though, Dr. Gregory L. Possehl, a Pennsylvania archeologist who specializes in the Indus civilization and had examined the pictures, cautioned against jumping to such conclusions. One had to be careful, he said, not to confuse potter's marks, graffiti and fingernail marks with symbols of nascent writing.

Of the earliest writing systems, scholars said, only the Sumerian, Chinese and Mesoamerican ones seemed clearly to be independent inventions. Reviewing the relationship between early Chinese bronze art, "oracle bones" and writing, Dr. Louisa Huber, a researcher at Harvard's Fairbanks Center for East Asian Research, concluded, "Chinese writing looks to be pristine."

But few pronouncements about early writing go undisputed. Dr. Victor H. Mair, a professor of Chinese language at Penn, offered evidence indicating, he said, that "the Chinese writing system may well have received vital inputs from West Asian and European systems of writing and protowriting."

Dr. Mair cited an intriguing correspondence between the Chinese script and 22 Phoenician letters and also Westernlike symbols on pottery and the bodies of mummies found in the western desert of China. The recent discoveries of the mummies, wearing garments of Western weaves and having Caucasoid facial features, have prompted theories of foreign influences on Chinese culture in the first and second millennia B.C. It had already been established that the chariot and bronze metallurgy reached China from the West.

Though no one seemed ready to endorse his thesis, Dr. Mair said, "We simply do not know for certain whether the Chinese script was or was not independently created."

Dr. Peter Damerow, a specialist in Sumerian cuneiform at the Max Planck Institute for the History of Science in Berlin, said, "Whatever the mutual influences of writing systems of different cultures may be, their great variety shows, at least, that the development of writing, once it is initiated, attains a considerable degree of independence and flexibility to adapt a coding system to specific characteristics of the language to be represented."

Not that he accepted the conventional view that writing necessarily started as some kind of representation of words by pictures. New studies of Sumerian protocuneiform, he said, challenge this interpretation. The structures of this earliest writing, for example, did not match the syntax of a language. Protocuneiform seemed severely restricted, compared with

When No One Read, Who Started to Write?

INDUS VALLEY

2500 B.C. Ceramic seals were used by the inhabitants of what is now Pakistan to indicate ownership or destinations for bundles of goods.

MESOAMERICA

250–300 B.C. The Olmecs, Mayans and Zapotecs are thought to be the first of many cultures in this region to develop writing.

SUMERIA

3200 B.C. Mesopotamian culture is generally thought to be the first culture to produce written texts. The Sumerians used a stylus and wet clay to record the ingredients for beer.

EGYPT

3200–3000 B.C. Early Egyptians used a system of pictorial hieroglyphics as a written language. The earliest examples were phonetically arranged symbols on small clay tablets used to indicate payments for commodities.

CHINA

1500–1200 B.C. During the Shang Dynasty the Chinese began to inscribe divinations on the underside of turtle shells and ox bones. These artifacts are now referred to as "oracle bone inscriptions."

spoken language, dealing mainly in lists and categories, not in sentences and narrative.

This presumably reflects writing's origins and first applications in economic administration in a growing, increasingly complex society, scholars said. Most of the Uruk tablets were documents about property, inventory and, even then, taxes. The only texts that do not concern administrative activities, Dr. Damerow said, were cuneiform lexicons that were apparently written as school exercises by scribes in training.

For at least two decades, in fact, Dr. Denise Schmandt-Besserat, a University of Texas archeologist, has argued that the first writing grew directly out of a counting system practiced by Sumerian accountants. They used molded clay tokens, each one specially shaped to represent a jar of oil, a large or small container of grain, or a particular kind of livestock. When the tokens were placed inside hollow clay spheres, the number and type of tokens inside were recorded on the ball with impressions resembling the tokens. Finally, simplifying matters, the token impressions were replaced with inscribed signs, and writing was invented.

Though Dr. Schmandt-Besserat has won wide support, some linguists question her thesis and other scholars, like Dr. Pittman of Penn, think it too narrow an interpretation. They emphasized that pictorial representation and writing evolved together, part of the same cultural context that fostered experimentation in communication through symbols.

"There's no question that the token system is a forerunner of writing, and really important," Dr. Pittman said in an interview. "But I have an argument with her evidence for a link between tokens and signs, and she doesn't open up the process to include picture making and all other kinds of information-storage practices that are as important as the tokens."

Dr. Schmandt-Besserat, who did not attend the symposium, vigorously defended herself in a telephone interview. "My colleagues say the signs on seals were a beginning of writing, but show me a single sign on a seal that becomes a sign in writing," she said. "They say that designs on pottery were a beginning of writing, but show me a single sign of writing you can trace back to a pot—it doesn't exist."

In its first 500 years, she asserted, cuneiform writing was used almost solely for recording economic information. "The first information that writing gives you is only the same information the tokens were dealing with," she said. "When you start putting more on the tablets, products plus the name of who has delivered and received them, that is where art would enter the picture. Then writing is out of the box, in all directions."

Dr. Damerow agreed that cuneiform writing appeared to have developed in two stages, first as a new but limited means of recording economic information, later as a broader encoding of spoken language for stories, arguments, descriptions or messages from one ruler to another.

Even so, it was a long way from the origin of writing to truly literate societies. At the symposium, scholars noted that the early rulers could not write or read; they relied on scribes for their messages, record keeping and storytelling. In Egypt, most early hieroglyphics were inscribed in places beyond the public eye, high on monuments or deep in tombs. In this case, said Dr. Pascal Vernus of the University of Paris, early writing was less administrative than sacred and ideological, "a way of creating and describing the world as a dominating elite wants it to be."

Dr. Piotr Michalowski, professor of Near East civilizations at the University of Michigan, said the Uruk protocuneiform writing, whatever its antecedents, was "so radically different as to be a complete break with the past, a system different from anything else." It no doubt served to store, preserve and communicate information, but also was a new instrument of power.

"Perhaps it's because I grew up in Stalinist Poland," Dr. Michalowski said, "but I say coercion and control were early writing's first important purpose, a new way to control how people live."

—JOHN NOBLE WILFORD, April 1999

New Family Tree Is Constructed for Indo-European

LINGUISTS HAVE LONG AGREED THAT languages as diverse as English, French, German and even Iranian, Albanian and Armenian all descended from a long-dead tongue, proto-Indo-European, spoken some 5,000 years ago. But the exact nature of the relationship has been a matter of dispute. How should the family tree of the Indo-European languages be drawn?

In what some linguists are welcoming as an important development in historical linguistics, researchers at the University of Pennsylvania have proposed an answer. Drawing on ideas in biology, computer science and historical linguistics, they have developed what they say is the most accurate picture yet of how the proto-Indo-European language fractured into the major branches that exist today.

The research by two linguists, Dr. Donald Ringe and Dr. Ann Taylor, and a computer scientist, Dr. Tandy Warnow, was described at the National Academy of Sciences Symposium on the Frontiers of Science in Irvine, California.

For simplicity's sake, linguists often depict the principal Indo-European language groups as 10 equal branches radiating like spokes from a wheel. But they do not seriously suppose that all these splits could have occurred simultaneously, as if 10 bands of adventurers had struck off in different directions from the Indo-European homeland.

More likely, one group parted ways, then another, which might have fractured into two more. When linguists sit down and calculate the possibilities, they are overwhelmed by a dizzying case of combinatorial explosion. By one reckoning, there are more than 34 million ways the genealogical chart for all the Indo-European languages could be drawn.

Using a new computer program, or algorithm, the scientists at the University of Pennsylvania have sifted through the myriad possibilities and come up with what they believe is the general shape of the tree. According to their picture, the first to split off from proto-Indo-European was Anatolian, an extinct group of languages, including ancient Hittite, which were once spoken in Turkey.

Next to split off were Celtic, which gave rise to Irish, Gaelic, Welsh and Breton, and Italic, the precursor to Latin and the Romance languages . These were followed by Armenian and Greek. Among the last to split off were the Indic and Iranian languages.

The exciting implication is that the tree shows, in schematic form, how the descendants of the original Indo-Europeans might have spread across Europe and Asia.

Consider, for example, the puzzling case of the ancient Germanics. Their language, which gave rise to English, Dutch and German, has stubbornly resisted efforts to fit it into the Indo-European tree. The new genealogy offers a possible reason: an early ancestor of the Germanic language was closely related to Balto-Slavic, whose modern descendants include Lithuanian, Latvian, Russian, Czech and Polish. Then its speakers migrated westward, coming into contact with speakers of Italic and Celtic.

"What really surprised us was that we were able to come up with a clear and consistent explanation for Germanic," Dr. Ringe said.

Dr. Craig Melchert, a linguist at the University of North Carolina, called the new technique "a promising innovation."

"Based on what I've seen so far," he said, "one undeniable merit of their model is that it allows for the testing of different hypotheses in a very explicit and efficient way not previously available."

But Dr. Melchert cautioned that the data fed into such a model were always open to interpretation and dispute. He, for one, strongly disagrees that the scientists have verified that Anatolian was the first to split away from proto-Indo-European. This theory, the Indo-Hittite hypothesis, has been debated for years.

Not everyone is convinced that computational studies have much to offer a science like linguistics, in which the data are so incomplete and indecisive. "What this emphasizes to me is that there are places for exact methods, and also places where it's silly to pretend to have them," said Dr. Jay Jasanoff, a Cornell University linguist. He said that in constructing such charts, there was always a danger that linguists would unconsciously pick data that confirmed their prejudices. "It's like 'scientific' poll taking," he said. "You put the questions in a way that gets the answer you're looking for."

But Dr. Ringe noted that before this study, he believed, contrary to his findings, that the Indo-Hittite hypothesis was wrong. "It might even be fair to say that I was biased against it," he said. "So you can imagine how startled I was when the algorithm kept turning up Anatolian as one first-order branch of the family, and everything else as the other first-order branch—exactly what the hypothesis says."

In tracing the pedigree of languages, linguists face the same problems biologists confront in drawing taxonomic charts of the species. Similar traits do not necessarily imply common descent; they might be coincidences. All birds are related because they have wings. But insects also have wings, as do bats. Try to connect all winged creatures into one big family and you would end up with a tangle instead of a tree. The similarity is an artifact of what biologists call "convergent evolution."

"Suppose two languages inherit the same word meaning 'winter,' and both of them independently shift its meaning so that it means 'snow' instead," Dr. Ringe said. "Greek, Armenian and Sanskrit actually did that. If we used our method blindly, we'd group those languages together because they all innovated in the same way, and we'd be wrong."

Even when linguists have good reason to believe that innovations are related, coming up with the right diagram can be a daunting task.

To constrain the explosion of possibilities, linguists need some way to sift out the best candidates. The most obvious criterion is Occam's razor: the simplest tree is most likely to be correct. But weeding out the most economical tree from the deluge of possibilities results in one of those problems that mathematicians call "intractable." As the size of the variables slowly increases, the computer time it takes to find the best tree soars. The problem would hopelessly overwhelm even the fastest conceivable computer.

By making artful assumptions, scientists can devise algorithms that will run in reasonable times. Dr. Ringe, Dr. Taylor and Dr. Warnow assumed that the best tree for the Indo-European language was what linguists and biologists call a "perfect phylogeny," with branches, limbs and twigs radiating neatly from the trunk without the tangles caused by accidents like convergent evolution.

Using an algorithm devised by Dr. Warnow and Dr. Sampath Kannan, they analyzed 12 languages representing the 10 major Indo-European branches. The computer grouped the languages according to whether they shared certain similarities.

In addition to cognates—related words with the same meaning, like English *mother*, German *Mutter* and Spanish *madre*—the algorithm searched for similar innovations in pronunciation. For example, "k" sounds at the beginning of Indo-European words become "h" sounds in the Germanic languages. While the initial consonant in the Latin word *cornu* (meaning "horn") is preserved in French (*corne*) and Spanish (*cuerno*), it shifted in ancient Gothic to *haurn* and is reflected in German as *Horn*, in Dutch as *hoorn* and in Norwegian and English as *horn*. Latin *centum* became the English *hundred* and German *hundert*. Latin *cor* and Greek *kardia* appear as the English *heart*, German *Herz*, Dutch *hart* and Norwegian *hjerte*.

The scientists also looked for similar grammatical innovations. Germanic languages, for example, share the same kind of past-tense marker (the "-ed" ending on regular English verbs). Nothing like this exists in other Indo-European languages.

When all the number crunching was done, the scientists were left with two diagrams showing possible relationships between the languages . But there were a number of points that could not be fit into a neat tree structure: 12 for one chart, 13 for the other. These anomalies were probably the result of chance similarities, like convergent evolution and undetected borrowings between languages. Since Germanic occupied very different places on the two trees, the linguists suspected that it was the culprit causing most of the problems.

Sure enough, when they removed Germanic from the pool, they came up with four much sturdier tree structures. This time each tree had only four or five anomalies, and the scientists were able to explain away all but two.

The first surprise was that in all four trees, the Anatolian language group immediately split away from proto-Indo-European, just as the Indo-Hittite hypothesis has held.

The second surprise was the evidence that Germanic had shifted allegiances, from Balto-Slavic to Italic and Celtic, like a country pulled between the orbits of neighboring powers. In the past, some linguists have used Germanic's ambiguous position in the Indo-European family to argue that, in their early stages, these languages were more like a network than a tree; instead of neatly cleaving one from the other and developing in isolation, they hovered nearby, trading innovations back and forth. But Dr. Ringe and his colleagues showed that Germanic might simply be a special case.

This conclusion has raised at least one pair of eyebrows. "It's a gratuitous interpretation," Dr. Jasanoff said. "The shifting position of Germanic could simply be interpreted as saying the method didn't work."

But Dr. Ringe argued, "We've shown what we principally set out to show, that this new computational method for evolutionary tree construction is more powerful and more reliable than any in existence, and that it is especially apt for the precise testing of hypotheses about the subgroupings of languages. We're hoping this will help lead to new directions of research in historical linguistics."

—GEORGE JOHNSON, January 1996

Scholars Debate Roots of Yiddish, Migration of Jews

TRYING TO TRACE THE ANCIENT roots of a modern language is always a maddeningly ambiguous and uncertain enterprise. With Yiddish, the language of the Ashkenazic Jews of Central and Eastern Europe, the task is even harder because of the horrifying fact that most of the speakers were exterminated in the Holocaust.

As a result, the study of Yiddish origins—and especially the touchy issue of its relationship to German—has sometimes been criticized as one in which rational analysis has been overwhelmed by emotion. But a number of studies are now being welcomed by linguists as evidence that the field is turning into a solid science.

"There are now signs that the history of Yiddish is becoming a scientific enterprise instead of the mythological exercise it used to be," said Dr. Jerrold Sadock, a linguist at the University of Chicago.

By trying to reconstruct the original Yiddish, linguists hope to explain the origins of this rich language, in which a largely Germanic grammar and vocabulary is mixed with Hebrew and Aramaic, and sprinkled with words from Slavic and ancient Romance languages . The question they hope to answer is whether Yiddish began in Western Europe and spread eastward, as the common wisdom holds—or whether, as an increasing number of scholars now believe, its origins lie farther east. One linguist has recently argued that Yiddish began as a Slavic language that was "relexified," with most of its vocabulary replaced with German words.

Arching over these questions is the central mystery of just where the Jews of Eastern Europe came from. Many historians believe that there were not nearly enough Jews in Western Europe to account for the huge population that later flourished in Poland, Lithuania, Ukraine and nearby areas.

By reconstructing the Yiddish mother tongue, linguists hope to plot the migration of the Jews and their language with a precision never possible before. It has even been suggested, on the basis of linguistic evidence, that the Jews of Eastern Europe were not predominantly part of the Diaspora from the Middle East, but were members of another ethnic group that adopted Judaism.

"Yiddish is widely perceived as a very special language, " said Dr. Alexis Manaster Ramer, a linguist at Wayne State University in Detroit. "If this is correct, the explanation might lie precisely in the historical uniqueness of the circumstances which produced Yiddish."

The revival of the field is due, in part, to a mammoth project at Columbia University to map the dialects of Yiddish, plotting precisely where on the European continent the many variations were once spoken. After decades of preparation, *The Language and Culture Atlas of Ashkenazic Jewry* began appearing in 1992, with volume one. The third installment was sent to the printers and is available from the publisher Max Niemeyer in Tubingen, Germany. At least seven more volumes are planned.

This accumulating evidence is being eagerly seized by linguists intent on tracing the roots of Yiddish. "The atlas is a fabulous tool for doing this kind of work," said Dr. Robert D. King, who holds the Audre and Bernard Rapoport Chair of Jewish Studies at the University of Texas at Austin. Work on the project began in the early 1960s after Dr. Uriel Weinreich of Columbia University and his wife, the folklorist Beatrice Silverman Weinreich, began an effort to interview some 600 Yiddish-speaking immigrants in Israel, the Alsace region of France, the United States, Canada and Mexico. When Dr. Weinreich died in 1967, the project was taken over by Dr. Marvin Herzog.

"The atlas is of monumental importance to the field of Yiddish studies," said Dr. Neil Jacobs, a linguist at Ohio State University in Columbus. The detailed interviews, each lasting some 15 hours and including more than 3,000 questions, provide an usually exact picture of both Yiddish dialects and culture. The atlas is so precise that it can show the line of demarcation separating Eastern European Jews who sugared their gefilte fish from those who did not, or between those who ate tomatoes and those who considered them *tref,* or "unclean," because of their bloodred color.

The linguistic information is just as precise—charting, for example, differences in the pronunciation of the word *flaysh,* or "flesh."

The emergence of this rich lode of information is expected to provide the kind of hard evidence that linguists need to separate hypothesis from speculation.

For centuries it was widely assumed that Yiddish was just broken German, more of a linguistic mishmash than a true language . Even the language's own speakers called it *Zhargon,* meaning "jargon." In the early 20th century, linguists found evidence that Yiddish and modern German were of equal stature—parallel offshoots of the same Germanic mother tongue. The other components of Yiddish were explained as superficial borrowings grafted onto an essentially Germanic language.

After the horrors of World War II, some Jewish scholars set out to distance Yiddish from German and show that it was a unique cultural creation of the Jews. The main champion of this view was Dr. Max Weinreich, the father of Uriel Weinreich and the driving force behind the YIVO Institute for Jewish Research, which began in Vilna, Lithuania, and is now in Manhattan. Noting that Yiddish includes a few words from Old Italian and Old French, Dr. Weinreich argued that it began as a Romance language that was later Germanized. In this view, Yiddish was invented by Jews who had arrived in Europe with the Roman army as traders, later settling in the Rhineland of western Germany and northern France. Mixing Hebrew, Aramaic and Romance with German, they produced a unique language, not just a dialect of German.

Pushed eastward by the religious zealotry arising from the medieval Crusades and the black plague, which fanatical Christians blamed on the Jews, the speakers of Yiddish reestablished themselves in Poland and surrounding areas, where the language picked up its Slavic content. According to this now dominant theory, there were very few Jews in Eastern Europe before the great immigration from the west. Yiddish is seen as a largely Western European phenomenon.

As appealing as this theory has been to Jews who wish to divorce the language from that of their Nazi persecutors, corroborating linguistic evidence has been sparse. Even more troublesome are demographic studies indicating that during the Middle Ages there were no more than 25,000 to

35,000 Jews in Western Europe. These figures are hard to reconcile with other studies showing that by the 17th century there were hundreds of thousands of Jews in Eastern Europe.

"You just can't get those numbers by natural population increase," Dr. King said. In a paper published in 1992, he argued that the origins of Yiddish were not in the Rhineland but eastward along the Danube—in Bavaria and as far east as Hungary and the Czech and Slovak lands. From there, he argues, the language radiated both westward, into the Rhineland, and eastward into Poland, Lithuania, Latvia and other areas.

Dr. King bases his conclusion on work he began in the 1980s with Dr. Alice Faber, a linguist now at Haskins Laboratories in New Haven, Connecticut. Dr. King and Dr. Faber found no significant similarities between the Yiddish of Eastern Europe and the dialects of German spoken in the Rhineland. They uncovered a few similarities between Yiddish and East Central German, spoken as far east as Poland. For example, German diphthongs like "ie" and "uo" were compressed in both languages, so that *knie* ("knee") was rendered *kni*. But the most striking resemblances were between Yiddish and Bavarian, a dialect of German. "Yiddish resembles nothing more closely than medieval Bavarian," Dr. King said.

For example, both Bavarian and Yiddish differ from German in that they have lost a pronunciation rule called "final devoicing." Germans pronounce *Tag* ("day") as though it ended in "k" and *Rad* ("wheel") as though it ended in "t." But in Yiddish and Bavarian the two words are pronounced *tog* and *rod*. Another example: the words *Blume* ("flower") and *Gasse* ("street") are pronounced with two syllables in German but with one syllable in Bavarian and Yiddish. Bavarian is the only major German dialect that, like Yiddish, has undergone these two kinds of transformations.

Dr. King concedes that a western origin for Yiddish is still possible: Jews migrating from the Rhineland may have lingered in the Danube region long enough for their language to significantly change. But he is skeptical that essentially all traces of Rhineland German could have been so completely erased.

Contrary to the common wisdom, Dr. King believes there must have already been a large population of Jews in Eastern Europe who had lived there since biblical times, coming up from the Middle East as traders speaking Hebrew and Aramaic. The Yiddish language and culture of the

Danube region then diffused eastward, he says, influencing this existing population.

Historians scarcely noticed these early pioneers, Dr. King speculates, because they did not have the leisure to develop the strong scholarly tradition that existed farther west. "The legacy of pre-Crusade Jewish life in Western Europe was a tradition of learning, of the rabbinate, of the community," Dr. King said. "The legacy of early Jewish life in the Slavic East was very largely the bones of its dead."

Some scholars believe the roots of Yiddish, and even the Ashkenazic people themselves, lie much farther east. In his 1976 book *The Thirteenth Tribe* Arthur Koestler made the startling suggestion, never taken seriously by linguists, that the Eastern European Jews were not really Semitic—that they were largely descended from the Turkish Khazars, who converted en masse to Judaism in medieval times.

More recently, Mr. Koestler's controversial thesis has been revived and expanded in a 1993 book, *The Ashkenazic Jews: A Slavo-Turkic People in Search of a Jewish Identity* (Slavica Publishers), by Dr. Paul Wexler, a Tel Aviv University linguist. Dr. Wexler uses a reconstruction of Yiddish to argue that it began as a Slavic language whose vocabulary was largely replaced with German words. Going even farther, he contends that the Ashkenazic Jews are predominantly converted Slavs and Turks who merged with a tiny population of Palestinian Jews from the Diaspora.

While few linguists are convinced by this radical hypothesis, the notion of a Slavic origin for Yiddish is being taken as a serious challenge to the field. "Even if he is not absolutely right," said Dr. Jacobs, "we are forced into a discussion of the issues he has raised."

In another reconstruction of proto-Yiddish, Dr. Manaster Ramer at Wayne State has uncovered evidence that some of Yiddish's Slavic words—like *nebbish,* referring to a pathetic individual—were part of the original language that grew into modern Yiddish. He has also found traces of western German dialects. But his analysis casts doubt on the hypothesis that Yiddish is an offshoot of Bavarian.

Dr. Manaster Ramer said that while traces of Bavarian were found in the Yiddish spoken in Eastern Europe, they did not show up in Western Yiddish, once spoken in western and southern Germany, the Netherlands, Switzerland and Alsace, or in medieval texts. He proposes that the Bavarian influence entered the language after Yiddish speakers had migrated eastward.

Linguists hope that in the next few years data like those gathered for the Columbia University atlas project will help them zero in on the Yiddish homeland.

—GEORGE JOHNSON, October 1996

Mummies, Textiles Offer Evidence of Europeans in Far East

IN THE FIRST MILLENNIUM A.D., people living at oases along the legendary Silk Road in what is now northwest China wrote in a language quite unlike any other in that part of the world. They used one form of the language in formal Buddhist writings, another for everyday religious and commercial affairs, including caravan passes.

Little was known of these desert people, and nothing of their language, until French and German explorers arrived on the scene at the start of this century. They discovered manuscripts in the now-extinct language,

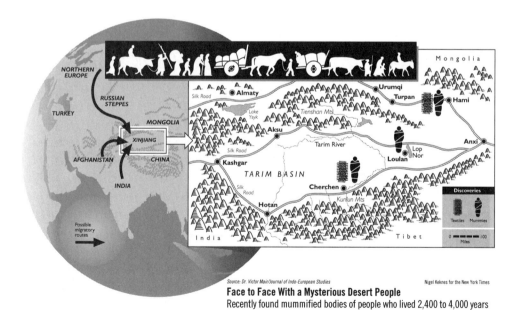

Face to Face With a Mysterious Desert People
Recently found mummified bodies of people who lived 2,400 to 4,000 years ago in the Tarim Basin region of Western China show them to have been strongly European in appearance, some resembling the Irish or Welsh. Their language, the now extinct Tocharian, also shows some similarities to Celtic and Germanic tongues.

which scholars called Tocharian and later were astonished to learn bore striking similarities to Celtic and Germanic tongues. How did a branch of the Indo-European family of languages come to be in use so long ago in such a distant and seemingly isolated enclave of the Eurasian landmass?

More surprises were in store. In the last two decades, Chinese archeologists digging in the same region, the Tarim Basin in Xinjiang Province, have uncovered more than 100 naturally mummified corpses of people who lived there 4,000 to 2,400 years ago. The bodies were amazingly well preserved by the arid climate, and archeologists could hardly believe what they saw. The long noses and skulls, blond or brown hair, thin lips and deep-set eyes of most of the corpses were all unmistakably Caucasian features—more specifically, European.

Who were these people? Could they be ancestors of the later inhabitants who had an Indo-European language? Where did these ancient people come from, and when? By reconstructing some of their history, could scholars finally identify the homeland of the original Indo-European speakers?

Linguists, archeologists, historians, molecular biologists and other scholars have joined forces in search of answers to these questions. They hope that the answers will yield a better understanding of the dynamics of Eurasian prehistory, the early interactions of distant cultures and the spread of kindred tongues that make up Indo-European, the family of languages spoken in nearly all of Europe, much of India and Pakistan and some other parts of Asia—and elsewhere in the world, as a result of Western colonialism.

At a three-day international conference, scholars shared their preliminary findings and hypotheses about how the Tocharian language and the Tarim Basin mummies might contribute to a solution to the Indo-European mysteries. The meeting, held at the University of Pennsylvania Museum, was organized by Dr. Victor H. Mair, a specialist in ancient Asian languages and cultures at the university. Some of the most recent research has been described in *The Journal of Indo-European Studies*.

Dr. Mair, who has spent several seasons in Xinjiang with groups studying the mummies and artifacts, said there was growing optimism that some important revelations might be at hand through genetic studies, a reinterpretation of ancient Chinese texts and art and a closer examination of textiles, pottery and bronze pieces.

"Because the Tarim Basin Caucasoid corpses are almost certainly the most easterly representatives of the Indo-European family and because they date from a time period that is early enough to have a bearing on the expansion of the Indo-European people from their homeland," Dr. Mair said, "it is thought they will play a crucial role in determining just where that might have been."

The tenor of discussions at the conference also reflected a critical philosophical shift that could affect attitudes toward other research problems in archeology and prehistory. Most participants invoked without apology the concept of cultural diffusion to explain many discoveries in the Tarim Basin.

For several decades, beginning in the 1960s, cultural diffusion was out of fashion as an explanation for affinities among widely scattered societies. The emphasis, instead, was on independent invention, and archeologists were often rebuked if they strayed from this new orthodoxy, which arose in part as a reaction to the political imperialism that often ignored or belittled the histories and accomplishments of subject lands. The Chinese, moreover, had long discouraged research on outside cultural influences, believing that the origins of their civilization had been entirely internal and independent.

But Dr. Michael Puett, a historian of East Asian civilizations at Harvard University, said the research on the Tocharians, the mummies and related artifacts revealed clear processes of diffusion. "Diffusionism needs to be taken seriously again," he said. Dr. Colin Renfrew, an influential archeologist at Cambridge University in England, made a point of endorsing this view.

Almost a century of studying the Tocharian manuscripts, dated between the sixth and eighth centuries A.D., has convinced linguists that the language represents an extremely early branching off the original, or proto-Indo-European, language. "That's the working hypothesis, at least for the moment," said Dr. Donald Ringe, a linguist at the University of Pennsylvania.

In that case, the people who came to speak Tocharian might have stemmed from one of the first groups to venture away from the Indo-European homeland, developing a daughter language in isolation. The fact that Tocharian in some respects resembles Celtic and Germanic languages does not necessarily mean that they split off together, scholars said, or that Tocharian speakers originated in Northern or Western Europe. Tocharian also shares features with Hittite, an extinct Indo-European language that was spoken in what is now Turkey.

One hypothesis gaining favor is that this scattering of Indo-European speakers began with the introduction of wheeled wagons, which gave these herders greater mobility. Working with Russian archeologists, Dr. David W. Anthony, an anthropologist at Hartwick College in Oneonta, New York, has discovered traces of wagon wheels in 5,000-year-old burial mounds on the steppes of southern Russia and Kazakhstan. Many scholars suspect that this region is the most likely candidate for the Indo-European homeland, though others argue for places considerably to the east or west or on the Anatolian plain of Turkey.

The possible importance of the wheel in Indo-European diffusion has been supported by evidence that wagons and chariots were introduced into China from the West. Wheels similar to those in use in western Asia and Europe in the third and second millenniums B.C. have been found in graves in the Gobi Desert, northeast of the Tarim Basin, and dated to the late second millennium B.C. Ritual horse burials similar to those in ancient Ukraine have been excavated in the Tarim Basin.

Linguists concede that their analyses of the ancient language will not produce answers to many of the questions about the Tocharians and their ancestors. Archeologists are more hopeful, with the mummy discoveries reviving their interest in the quest.

Early in this century, explorers and archeologists turned up a few mummies in the sands of China's western desert. One reminded them of a Welsh or Irish woman, and another reminded them of a Bohemian burgher. But these mummies, not much more than 2,000 years old, were dismissed as the bodies of isolated Europeans who had happened to stray into the territory and so were of no cultural or historical significance.

But no one could ignore the more recent mummy excavations, from cemeteries ranging over a distance of 500 miles. Not only were they well preserved and from an earlier time, but the mummies were also splendidly attired in colorful robes, trousers, boots, stockings, coats and hats. Some of the hats were conical, like a witch's hat. The grave goods included few weapons and little evidence of social stratification. Could this have been a relatively peaceful and egalitarian society?

One of the most successful excavators of mummies is Dr. Dolkun Kamberi, a visiting scholar at the University of Pennsylvania. He is a member of the dominant Turkic-speaking Muslim ethnic group in the Tarim Basin today, the Uighurs (pronounced WE-gurs). They moved to the area in the eighth century, supplanting the Tocharians, though Dr. Kamberi's fair skin and light brown hair suggests a mixing of Tocharian and Uighur genes.

His most unforgettable discovery, Dr. Kamberi said, came in 1985 at Cherchen on the southern edge of the Taklamakan Desert, an especially forbidding part of the Tarim Basin. The site included several hundred tombs in the salty, sandy terrain. In one tomb, he found the mummified corpse of an infant, probably no more than three months old at death, wrapped in brown wool and with its eyes covered with small flat stones. Next to the head was a drinking cup made from a bovine horn and an ancient "baby bottle," made from a sheep's teat that had been cut and sewn so it could hold milk.

In a larger tomb, Dr. Kamberi came upon the corpses of three women and one man. The man, about 55 years old at death, was about six feet tall and had yellowish brown hair that was turning white. One of the better-preserved women was close to six feet tall, with yellowish-brown hair dressed in braids. Both were decorated with traces of ocher facial makeup.

Among the other sites of mummy discoveries are cemeteries at Loulan, near the seasonal lake of Lop Nor and outside the modern city of Hami. Dr. Han Kangxin, a physical anthropologist at the Institute of Archeology in Beijing, has examined nearly all the mummies and many other skulls. At the Lop Nor site, he determined that the skulls were definitely of a European type and that some had what appeared to be Nordic features. At Loulan, he observed that the skulls and mummies were primarily Caucasian, though more closely related to Indo-Afghan types.

In nearly all cases, Dr. Han concluded, the earliest inhabitants of the region were almost exclusively Caucasian; only later do mummies and skulls with Mongoloid features begin to show up. At Hami, Caucasian and Mongoloid individuals shared the same burial ground and, judging by their dress and grave goods, many of the same customs.

Scientists have so far been permitted to conduct genetic studies on only one sample, from a 3,200-year-old Hami mummy. Although the recovered DNA samples were badly degraded, Dr. Paolo Francalacci of the University of Sassari in Italy said that he had been able to determine that the individual had belonged to an ancient European genetic group. He emphasized that the findings were preliminary.

"You can look at the mummy and see it's Caucasoid," Dr. Mair said. "Now we have genetic evidence. This is an important moment in our research."

The graves at Cherchen and Hami also produced the most intriguing textile samples from the late second millennium B.C. One of the Hami fragments was a wool twill woven with a plaid design, which required looms that had never before been associated with China or eastern Central Asia at such an early date. Irene Good, a specialist in textile archeology at the Pennsylvania museum, said that the plaid fabric was "virtually identical stylistically and technically to textile fragments" found in Austria and Germany at sites from a somewhat later period, about 700 B.C.

Dr. Elizabeth J. W. Barber, a linguist and archeologist at Occidental College in Los Angeles and the author of *Prehistoric Textiles* (Princeton University Press, 1991), said that plaid twills had first been discovered in the ruins of Troy, from about 2600 B.C., but had not been common in the Bronze Age. "My impression," she said, "is that weavers from the West came into the Tarim Basin in two waves, first from the west in the early second millennium B.C. and then from the north several centuries later."

Other evidence also seems to point to multiple ancient migrations into the Tarim Basin. "While it is clear that the early inhabitants of the Tarim Basin were primarily Caucasoids," Dr. Mair has written, "it is equally clear that they did not all belong to a single homogeneous group. Rather, they represent a variety of peoples who seem to have connections with many far-flung parts of the Eurasian land mass for more than two millennia."

Whoever they were, scholars said, many of the earlier mummified people were probably ancestors of the Tocharian speakers of the first millennium A.D. But no one knows if those early people spoke Tocharian. As Dr. James Patrick Mallory, an archeologist at Queen's University in Belfast, Northern Ireland, remarked, the mummies did not die "with letters in their pockets."

With much arm waving in front of maps, scholars speculated on the routes that Indo-European speakers might have followed into the Tarim Basin. Perhaps the earliest migrants, who looked most like Europeans, arrived from the north and northwest, over the mountains from Siberia or Russia. Later migrants, Caucasians but with Indo-Iranian affinities, could have moved in from the west and southwest.

After reviewing the many migration theories, including just about everything short of prehistoric parachute drops, Dr. Mallory sensed the audience's growing perplexity. "If you are not confused now, you have not been paying attention," he said.

One thing seemed clear to the scholars, however. Though East may be East, and West may be West, the twain met often in early times and in places, like the bleak Tarim Basin, that would have surprised Kipling. And these meetings did not begin with the Silk Road, the transcontinental trade route that history books usually describe as opening in the second century B.C. There never was a time, Dr. Mair said, "when people were not traveling back and forth across the whole of Eurasia."

Scholars doubt that these early movements usually took the form of mass migrations over long distances. But after the introduction of wheeled vehicles, pastoral societies could have begun extending their range over generations, coming into contact with others and finding more promising niches far from their linguistic origins.

"For people not in a hurry," said Dr. Denis Sinor, a historian at Indiana University in Bloomington and editor of *The Cambridge History of Early Inner Asia,* "the Eurasian continent was a very small world indeed."

In an article on the research, Dr. Mair wrote, "I do not contend that there were necessarily direct links stretching all the way from northwest

Europe to southeast Asia and from northeast Asia to the Mediterranean, but I do believe that there is a growing mountain of hard evidence which indicates indubitably that the whole of Eurasia was culturally and technologically interconnected."

—John Noble Wilford, May 1996

Luigi Luca Cavalli-Sforza;
A Geneticist Maps Ancient Migrations

THE GENES OF MODERN POPULATIONS carry the encoded history of humans' remote past and their early wanderings around the globe. Deciphering that history has been the preoccupation of population geneticists for a century, but key pieces of information have begun to fall quickly into place only in the last 10 years, since the advent of rapid methods for working out the chemical sequences of genes.

Standing in the midst of this giant jigsaw puzzle is a soft-spoken, Italian-born scientist with the mind of a mathematician and the interests of a philosopher: Dr. Luigi Luca Cavalli-Sforza. Perhaps more than anyone else in his field, Dr. Cavalli-Sforza, a 71-year-old genetics professor at Stanford University Medical School, has been able to make sense of the whisperings of human ancestors that are recorded in the genes of present-day people. His graphic maps make evident at a glance the patterns of ancient migrations, and they hint at some of the events that may have prompted them.

Dr. Cavalli-Sforza and two coauthors, Dr. Paolo Menozzi and Dr. Alberto Piazza, have written a book that has taken 12 years to produce: a genetic atlas called *The History and Geography of Human Genes,* published by Princeton University Press. The authors are also the movers behind the Human Genome Diversity Project, an ambitious effort to arrange for the collection of DNA samples and anthropological information from 25 individuals in each of 400 targeted populations around the world.

It will be the first time such information has been systematically collected from so many populations and should allow geneticists to reconstruct the history of the world's populations in detail. Another aim of the project is to prepare the collected DNA samples so they can be easily replicated, thus preserving a genetic record of the world's diverse populations.

In the United States, the Human Genome Diversity Project has received start-up financing from the National Science Foundation, the National Institutes of Health and the Department of Energy. Although the project is barely off the ground and future financing is far from certain, Dr. Cavalli-Sforza and his staff are preparing computer programs for the mountains of data the new research is expected to generate. One such program, dubbed "genography," has summarized genetic and linguistic data on close to 1,000 population studies done by scientists all over the world from 1961 to 1989.

The story emerging from Dr. Cavalli-Sforza's studies, as reported in the journal *Science,* confirms and extends the findings from other fields of inquiry into human origins. The genetic studies bear witness to two major phases of human expansion: the first was the movement of *Homo erectus* out of Africa one million to two million years ago; the second was the spread of *Homo sapiens* around the globe, beginning 100,000 years ago and reaching every continent 60,000 years ago.

Because the genetic makeup of African populations today displays such a wide diversity in the DNA of mitochondria, cellular structures that supply energy and are inherited solely through the mother's line, Dr. Cavalli-Sforza believes this second wave of migration also originated in Africa. At this time, the human brain reached its present size, and he contends that language and the use of boats and rafts, common about 55,000 to 60,000 years ago, supported the spread of populations into every continent, even Australia. After this primary phase, local expansion patterns become more complex and more difficult to decipher, he says.

By analyzing patterns of geographic variation of genes and their relative importance, Dr. Cavalli-Sforza has identified several strong patterns of expansion in Europe. In general, in choosing contemporary populations to study, he has concentrated on aboriginal people who were in place before the great age of European colonial expansion in the late 15th and 16th centuries and has avoided modern cities, where the recent intermingling of peoples tends to obscure the genetic record.

Confirming the general picture pieced together by archeologists and linguists, he finds that after the introduction of agriculture in the Middle East about 10,000 years ago, farmers from there spread at a rate of about one kilometer, or five eighths of a mile, a year, eventually settling throughout Europe. These early farmers replaced the nomadic hunter-gatherers, and their sole direct survivors are believed to be the Basques, who are genetically and linguistically far removed from other Europeans.

Dr. Cavalli-Sforza's work has also given a genetic foundation for a controversial theory of Dr. Marija Gimbutas, of the University of California at Los Angeles, that between 6,000 and 4,000 years ago, an expansion of people began in the area of the southwest Russian steppes, where the domestication of the horse took place. The people in these regions migrated very rapidly across Europe, spreading their language, which is the common mother tongue from which all European languages derived, the theory says.

"We discovered an area of population expansion that almost perfectly matched Gimbutas's projection for the center of Kurgan culture," Dr. Cavalli-Sforza said.

In Africa, population movements again followed agricultural development. Before the Sahara became a desert, 3,000 to 4,000 years ago, it was the center of farming and cattle breeding in Africa. When the climate changed there, the farmers moved first to western Africa, where sorghum and millet began to be cultivated, and then the pastoralists began to migrate south and east.

Although archeological evidence points to Mexico as the seat for agricultural development in North America about 5,000 years ago, Dr. Cavalli-Sforza has as yet found no genetic pattern of population movement emanating from Mexico at that time, as might be expected. It is not surprising, however, that there is a clear pattern of European settlement beginning on the east coast and spreading west, and three genetically distinct groups of native North Americans: Eskimos, the Na-Dene in Canada and American Indians.

Native North Americans fall into three distinct genetic patterns, Dr. Cavalli-Sforza found, supporting the hypothesis of the linguist Dr. Joseph Greenberg, retired from Stanford University, that there were three separate migrations of Asians across the Bering Strait. Dr. Cavalli-Sforza's genetic data suggest that the earliest migrants were American Indians and that they arrived here about 30,000 years ago.

Dr. Cavalli-Sforza, who was born in Genoa, Italy, began his genetic research with bacteria after receiving a medical degree from the University of

Pavia in 1944. In 1948, he worked with Sir Ronald A. Fisher, one of the early pioneers in population genetics, at Cambridge University in England.

"I've always been interested in quantitative prediction," he said in an interview in his office. "I probably should have been a physicist."

In the 1950s, while lecturing at the University of Parma and the University of Pavia, he studied the populations of the small towns in the Parma Valley in northern Italy. He discovered that the smaller the village, the greater the variation between families because the small number of individuals exaggerated the impact of random genetic variations, in this case albinism, mental deficiency and deaf-mutism. The study was one of the first to document "genetic drift," genetic changes that have nothing to do with Darwin's law of natural selection or "survival of the fittest." Certain changes, he found, just happened.

"It wasn't that Darwin was wrong; there was just something extra," he said. "Motoo Kimura calls it 'the survival of the luckiest' and believes that genetic drift is as important a force in evolution as natural selection." Dr. Kimura works at the National Institute of Genetics in Japan.

In the 1960s, Dr. Cavalli-Sforza began working with Dr. Anthony Edwards at the Institute of Genetics at the University of Pavia on a method for translating the jumble of data on genetic variation in populations. Together they presented an important paper, "Analysis of Human Evolution," at the International Congress of Genetics in The Hague in 1963, outlining their idea of organizing data into components based on greater and smaller percentages of variation.

Thus isolated, each component variable could be graphed and analyzed, and the "genetic distance" that separated two populations precisely measured. For this initial study, Dr. Edwards and Dr. Cavalli-Sforza tested 15 populations, three on each of five continents—Australia, Europe, Asia and North and South America—for the five major blood groups.

Only in the last 10 years have Dr. Cavalli-Sforza and his collaborators begun to look at DNA markers to measure the amount of inherited genetic variation. One of the most useful markers has proved to be in short repeated segments of DNA known as "microsatellites," which appear in every chromosome. The microsatellites, whose purpose is unknown, are so variable as to serve like fingerprints in identifying individuals.

In a study by Dr. Cavalli-Sforza and Dr. Anne M. Bowcock of the University of Texas Southwestern Medical Center in Dallas, in which 30 different microsatellites were analyzed in 148 individuals around the world, 87 percent of the individuals could be identified as to their place of origin just by looking at the lengths of short repeats of nucleotides, the chemical units that make up DNA. In the case of Pacific Basin populations, the microsatellites indicated precisely what island cluster they came from.

The Armed Forces Institute of Pathology, which keeps a registry of DNA samples from active military personnel, is developing systems using about a dozen microsatellites to identify soldiers lost in combat. In the future, any fragment of material containing a good sample of DNA could allow an individual to be identified.

Dr. Cavalli-Sforza's genetic studies using DNA markers have in many cases confirmed the earlier studies based on blood types, but the new studies have also indicated that Europeans are a mixed population that emerged only about 30,000 years ago and appears to have 65 percent Asian ancestry and 35 percent African ancestry (with an error rate of plus or minus 8 percent). Australian aboriginals, though they appear to look more like Africans, are genetically closer to Chinese. All races or ethnic groups now seem to be a bewildering array of overlapping sets and subsets that are in a constant state of flux in fairly short periods of evolutionary time.

At a hearing of the Senate Committee on Government Affairs on the Human Genome Diversity Project, Dr. Cavalli-Sforza and Dr. Mary-Claire King, a geneticist at the University of California at Berkeley, discussed the implications of their work. They called racism "an ancient scourge of humanity" and expressed the hope that the extensive study of world populations would "undercut conventional notions of race and underscore the common bonds between all humans."

—LOUISE LEVATHES, July 1993

2
LANGUAGE IN OTHER SPECIES

Humans are the only living species with a spoken language. The Neanderthals and other extinct species of hominid may have spoken, but there is no direct evidence that they did.

It is a little surprising for evolution suddenly to produce a totally new trick from its hat; gradual change is its usual mode of operation. So biologists have naturally searched for the rudiments of language in other animals.

On close analysis, other species do turn out to possess remarkable communication systems. Ants have an elaborate system of chemical signals that guide their social behavior. Vervet monkeys make one kind of warning noise when they spot an eagle, another kind if they see a snake.

Given these hopeful antecedents, a series of researchers have explored chimpanzees' capacity for language. As our closest living relatives, they if any species should know how to communicate.

Chimpanzees do not have the vocal apparatus for speech, so researchers have tried instead to teach them various forms of symbolic language, whether Ameslan, the sign language used by the deaf, or symbols activated through a computer keyboard.

Chimps are highly intelligent animals and can clearly learn the meaning of individual symbols. They can also string several symbols together, often in highly evocative ways. But the essence of language is syntax, the rules for combining words in a sentence. Most linguists remain to be persuaded that chimpanzees have or can learn true syntax.

Look Who's Talking.
Don't Bother Listening.

SO ANCESTORS OF HUMAN BEINGS might have had the capability to speak as early as 400,000 years ago, as anthropologists just reported. But what did they have to say? Probably much more than an occasional yabba dabba doo.

They may have huddled around an open fire and cursed Og for letting a choice mammoth get away. Or they shivered in a rock shelter and talked over plans to move south for the winter, following the game. Even then, weather was a conversational standby: My, my, aren't winters colder now than in olden times?

Anthropologists suspect there was something else familiar about ancestral chitchat. They probably gossiped a lot about family, social relationships, sickness and death. If that poor young mother dies, who will take care of her baby? Is the head of the clan the hunter he once was? Is your son interested in my daughter? Perhaps Og was mooning over that woman in the neighboring clan when he should have had his mind on that mammoth.

What the earliest speaking ancestors spoke about is hardly a frivolous question. It may reflect the evolutionary root of language and speech, behavior critical in setting humans apart from other animals.

What they said, though, is less important to scientists than when they actually said it. The scientists at Duke University announced that they explored a new avenue of fossil anatomy and found surprising evidence suggesting that these vocal abilities may have evolved earlier than previously thought. The much-maligned Neanderthal could probably speak, though perhaps not well enough to ward off extinction 30,000 years ago.

"When you look around us, only one animal has language and speech," said Dr. Matt Cartmill, an anthropologist at Duke University, whose work with Dr. Richard F. Kay and a former student, Michelle Balow, was published in *The Proceedings of the National Academy of Sciences*. "That permits us to have a power over the world that no other animal has."

Ancestors as far back as *Homo erectus* more than a million years ago had developed a brain about 80 percent the size of the human brain and probably had a lot on their minds before they had the vocal equipment to utter their thoughts and feelings. They were beginning to live in larger groups, sharing work and responsibilities, dealing with increasingly complex social relations by imposing symbolic meaning on reality. Their brains—scientists see a correlation between brain size and the size of living groups—thus could have been capable of language before they could speak a word.

The connection between social relations and speech is the reason Dr. Robin Dunbar, a British anthropologist, thinks it very likely that primeval conversation was spiced with gossip, antecedents of the kaffeeklatsch and talk shows.

The research by the Duke physical anthropologists has revived debate about the origin and timing of human speech. Whether *Homo erectus*, a significant transition species between primitive and more recent human ancestors, could speak is problematic. The earliest unequivocal evidence for human speech is the cave art and other artifacts of modern *Homo sapiens*, beginning some 40,000 years ago. Imagine the critics coming and going at Lascaux and speaking of some cave Michelangelo.

In any event, the cave painters left the first clear expressions of symbolic thought and a strong argument for speech as a recent development.

But the Duke researchers found anatomical evidence suggesting that vocal capabilities like those of modern humans may have evolved in archaic *Homo sapiens*, including Neanderthals. They compared the diameter of a hole at the bottom of the skulls in modern humans, apes and several earlier species of the genus *Homo*.

The hole is the hypoglossal canal through which nerve fibers from the brain pass to control muscles of the tongue. Measurements showed that the passage in modern human skulls is twice the width of those in speechless chimpanzees and some of the more distant human ancestors. But in 400,000-year-old fossils of the *Homo* line, the hypoglossal canals fell within the size range of modern humans and thus could have carried enough nerves for tongues to form speech sounds. It is only suggestive evidence, not proof, but enough to intrigue other scientists.

One problem, Dr. Cartmill said, is sorting out the physical attributes crucial to speech. The position of the larynx, the voice box, is undoubtedly important. In apes it sits high in the neck. It is lower in humans, which facilitates the utterance of a wide range of sounds but makes people the only mammals incapable of simultaneously drinking and breathing.

If the content of ancestral talk was familiar, the sounds may have been less so, outside the nursery. Infants are born with a high larynx, which limits their ability to form any but the simplest word sounds like goo, dada, mama. Only when the larynx descends do they develop fluency. Perhaps, anthropologists say, early ancestors were similarly restricted. Vowel sounds like "e," "a" and "o" are characteristic of all languages today, suggesting deep origins.

Dr. Philip Lieberman, a cognitive scientist at Brown University, who wrote *Eve Spoke* (Norton, 1998), imagines early speech as short, simple words strung together in short phrases or sentences uttered slowly. Studying children, he noted that they talk about 50 percent more slowly than adults until about the age of 10. It is "a reasonable guess," he said, that the same applied to early ancestors.

Dr. Ian Tattersall, an evolutionary biologist at the American Museum of Natural History in New York City and author of *Becoming Human* (Harcourt and Brace, 1998), also looks to children as possible models of how language and speech developed. At first, their sounds are elemental, crying, laughing and grunting. With growth and experience, undifferentiated connections in their brains are "rewired" with new pathways permitting speech of increasing complexity. This may be one reason, he said, that learning a new language becomes much more difficult after the age of 10.

Whoever had the first word, even the Duke scientists realize theirs will not be the last word on the origin of human speech.

—JOHN NOBLE WILFORD, May 1998

Ancestral Humans Could Speak, Anthropologists' Finding Suggests

WHILE SCIENTISTS AGREE THAT SPEECH is probably the most important behavioral attribute that distinguishes human beings from other animals, they have been at a loss to determine when and how that transforming evolutionary step occurred.

They have probed the human brain and compared it with casts of the braincase from ancient fossil skulls. They have compared bones and muscle attachment points in the throats of humans, apes and ancestral human skeletons. Archeologists have examined patterns in early stone tools for clues to when humans might have developed the creativity and the self-awareness usually associated with communication skills like speech.

All they had been able to agree on is that the earliest unambiguous evidence for human speech is found in the cave art and other artifacts, particularly in Europe and Africa, that began appearing some 40,000 years ago.

Scientists at Duke University have explored a new avenue of fossil anatomy and found surprising evidence suggesting that vocal capabilities like those of modern humans may have evolved among species of the *Homo* line more than 400,000 years ago.

By then, their research shows, human ancestors may have had a full modern complement of the nerves leading to the muscles of the tongue and so could have been capable of forming speech sounds.

These findings, moreover, indicate that the Neanderthals, relatives of modern humans, could have had the same gift for speech. Their extinction about 30,000 years ago has often been attributed in part to speech deficiencies, restricting their ability for cultural innovation.

In a report published in *The Proceedings of the National Academy of Sciences,* the Duke anthropologists say that if their interpretation involving the tongue nerves is correct, "then humanlike speech capabilities may have evolved much earlier than has been inferred from the archeological evidence for the antiquity of symbolic thought."

The research was conducted by Dr. Richard F. Kay and Dr. Matt Cartmill at the Duke Medical Center in Durham, North Carolina, with the assistance of a former student, Michelle Balow. The results were also described in Salt Lake City at a meeting of the American Association of Physical Anthropology.

"This is evidence for the proposition that Neanderthals could talk," Dr. Cartmill said in a telephone interview. "Did they sound like modern humans? I don't know."

Anthropologists familiar with the research said the findings were interesting and exciting. Some were reserving judgment, but not Dr. Erik Trinkaus, an anthropologist at Washington University in St. Louis, who specializes in Neanderthal studies.

"I think it's not only a reasonable conclusion," he said, "but one long overdue."

Dr. Trinkaus said previous research had been based on deficient anatomical reconstructions, none of which adequately took into account the neurological aspects for controlling the vocal track to allow for speech. As for the possibility of speech by archaic *Homo sapiens* 400,000 years ago, even before Neanderthals, he said this was consistent with a significant enlargement of brain size in that period, the appearance of a more complex tool technology and migrations into colder climates, where life probably depended on greater planning that could be related to advances in communications skills.

On the other side, Dr. Philip Lieberman of Brown University, an authority on early language, has argued that the Neanderthal throat would not have been well suited for the production of the vowels a, i and u. But Dr. Trinkaus contended that a species would not have needed to produce all those sounds in order to have speech and language.

Even the discovery in Israel a decade ago of a Neanderthal skeleton with a large hyoid bone, which is in the throat and associated with speech, had not settled the issue of Neanderthal speech. Scientists had said there was still insufficient fossil evidence to enable an understanding of how the large hyoid bone might have influenced the production of vocalizations.

Dr. Cartmill himself cautioned that the new evidence for earlier human speech "is suggestive, but, in the present state of our knowledge, it is not proof."

Other scientists noted that other, independent evolutionary developments, including a lengthened larynx, enlarged prefrontal brain lobes and some reconfigurations of the brain, would have been critical to the emergence of speech. The size of the brain of Neanderthals was well within the range of that of modern humans.

The Duke scientists directed their research at the hypoglossal canal, a hole at the bottom of the skull in the back, where the spinal cord connects to the brain. Through the canal run nerve fibers from the brain to the muscles of the tongue.

It occurred to the scientists that the size of the hypoglossal canal might serve as an index of the vocal abilities of modern and early humans. The wider the canal, they assumed, the more nerve fibers there could be to control the tongue muscles. And the more nerves, they suggested, the finer control the species could have over its tongue for the purpose of making speech sounds.

The researchers compared measurements of hypoglossal canals of modern humans, apes and several human ancestor fossils, and concluded that the canals of modern humans are almost twice as large as those of modern apes—the chimpanzee and the gorilla—which are incapable of speech. They also found that the canal size of australopithecines, earlier human relatives that died out about one million years ago, did not differ much from that of chimpanzees.

The results, the scientists reported, "suggest minimum and maximum dates for the appearance of the modern human pattern of tongue motor innervation and speech abilities."

To narrow the range, the scientists examined skeletons of Neanderthals and also of species of the *Homo* genus that lived as much as 400,000 years ago. These included Kabwe specimens from Africa and Swanscombe fossils from Europe. Their hypoglossal canals fell within the range of those of modern *Homo sapiens*.

"By the time we get to the Kabwe, about four hundred thousand years ago, you get a canal that's a modern size," Dr. Cartmill said. "And that's true of all later *Homo* species, including Neanderthal."

—John Noble Wilford, April 1998

Brain of Chimpanzee Sheds Light on Mystery of Language

A SURPRISING STUDY HAS REVEALED that chimpanzees have a structure in their brains that is similar to a so-called "language center" in human brains, challenging cherished notions of how language evolved in humans and why apes cannot talk.

In most people, the structure, a slender inch-long piece of tissue called the planum temporale, is larger in the left side of the brain than the right. Since this area is involved in the processing and comprehension of speech sounds and sign language, scientists concluded 30 years ago that an enlarged plenum temporale in the left hemisphere was required for language and may have evolved for this purpose. Until now, no other animal was shown to have the same asymmetry in this brain region, located at the side of the head and connected to the ears.

The study, published in the journal *Science,* was carried out by three anthropologists—Dr. Patrick Gannon of the Mount Sinai School of Medicine in New York, Dr. Ralph Holloway of Columbia University, Dr. Douglas Broadfield of the City University of New York—and Dr. Allen Braun, a neurologist at the National Institute on Deafness and Other Communication Disorders in Bethesda, Maryland.

"This is an interesting and useful finding," said Dr. Antonio Damasio, an expert on brain and language at the University of Iowa College of Medicine in Iowa City. It shows the dangers of concentrating on brain centers and areas, he said, and supports research showing that language is widely distributed in the brain and probably evolved from novel connections rather than from new structures.

Language evolution remains a profound mystery. Sometime within the last two million years, two-legged primates, or hominids, developed the ability to talk with words, a dazzlingly difficult skill, said Dr. Terrence Deacon, a biological anthropologist at Boston University and McLean Hospital at Harvard Medical School.

Language requires lightning-fast processing of speech and understanding of abstract symbols—traits that other animals seem not to possess to the degree that humans do, he said. One can imagine the brain undergoing important reorganizations that underlie the ability to argue, cajole, complain and pontificate.

In the late 1960s, scientists were strongly influenced by the idea that bigger is better and that evolutionary pressures would lead to obvious structural changes in the human brain, Dr. Deacon said. If they could find unique bits of human brain architecture, they concluded, it might explain language. A study conducted in 1968 seemed to confirm this view, Dr. Deacon said. Of 100 human brains examined, 68 had an enlarged left planum temporale, 24 had structures of equal size and 11 had larger plana temporale on the right side.

Many people took this to mean that the planum temporale might be a "control center" for language, Dr. Gannon said. It is part of the auditory association cortex where sounds come in from the ear, are processed and sent to other parts of the brain. Further evidence stemmed from links between the planum temporale and "a melange of behaviors and disorders including musical talent, handedness and schizophrenia," Dr. Gannon said.

A few years ago, Dr. Gannon and his colleagues were examining preserved chimpanzee brains with the same methods used in the 1968 study. "We were simply exploring, looking for asymmetries, when one day our eyes popped out," he said. Of 18 chimpanzee brains examined, 17 had enlarged plana temporale on the left side of the brain. "This was more pronounced than in humans," Dr. Gannon said in a telephone interview.

Because chimpanzees cannot talk or play the violin, what does the finding mean? Aside from the obvious fact that the common ancestor of chimpanzees and humans had this brain asymmetry eight million ago, Dr. Gannon said there are three possibilities.

First, the asymmetry in the common ancestor is unrelated to language or communication. But later on, humans built on it and evolved the unique capacity for language. The planum temporale in chimpanzees did not evolve along the same path and plays an unknown role.

Second, the ancestral planum temporale is involved with communication but followed different trajectories in the two species. In humans it laid the basis for spoken and sign language and in chimps it laid the basis for a more gesture-based language.

"Chimps may have their own sophisticated form of language that we fail to recognize," Dr. Gannon said. "They have sense of self, can deceive one another, and show many complex communicative behaviors. Our language is vocal and auditory. Their language is gestural and visual."

The third possibility is that the planum temporale is not directly related to language or communication but has tangential functions, and its role in language has been vastly overrated.

"I think this study provides a strong demonstration that this particular brain asymmetry is not likely to be crucial for language, " said Dr. Deacon, whose book *The Symbolic Species* (Norton, 1997) lays out modern theories of language evolution. It supports the idea that humans did not evolve new brain structures for language but used structures that were present in other animals, he said.

To find out how language really evolved, researchers are looking more at microcircuitry than at gross anatomy, Dr. Deacon said. After all, 30 of every 100 people on average do not show the asymmetry, yet they appear to use language just like everyone else.

Studies show that there is tremendous variability in where language ends up in each person's brain, he said, and it can even move around in young adulthood after injury or, as one study showed, in learning how to do simultaneous translations. In that study, one language stayed on the left and the second language literally shifted to the right side of brain.

Better answers about language evolution lie in the way regions are connected in the brain, said Dr. Jeffrey Hutsler, a research assistant professor at Dartmouth College in Hanover, New Hampshire, who dissects human brains to look for such clues. Patches of connected cells in so-called language areas are laid out differently in the left and right sides of the brain, he said. Such structural variations could lead to different firing rates among cells, making some better at processing fast speech sounds.

In the meantime, no one has a clue about the function of the large left planum temporale in chimpanzees. They may use it for hearing calls and hoots and other sounds, Dr. Gannon said, or they may have traits that are on the threshold of human abilities.

—SANDRA BLAKESLEE, **January 1998**

Chimp Talk Debate: Is It Really Language?

PANBANISHA, A BONOBO CHIMPANZEE WHO has become something of a star among animal language researchers, was strolling through the Georgia woods with a group of her fellow primates—scientists at the Language Research Center at Georgia State University in Atlanta. Suddenly, the chimp pulled one of them aside. Grabbing a special keyboard of the kind used to teach severely retarded children to communicate, she repeatedly pressed three symbols—"Fight," "Mad," "Austin"—in various combinations.

Austin is the name of another chimpanzee at the center. Dr. Sue Savage-Rumbaugh, one of Panbanisha's trainers, asked, "Was there a fight at Austin's house?"

"Waa, waa, waa" said the chimpanzee, in what Dr. Savage-Rumbaugh took as a sign of affirmation. She rushed to the building where Austin lives and learned that earlier in the day two of the chimps there, a mother and her son, had fought over which got to play with a computer and joystick used as part of the training program. The son had bitten his mother, causing a ruckus that, Dr. Savage-Rumbaugh surmised, had been overheard by Panbanisha, who lived in another building about 200 feet away. As Dr. Savage-Rumbaugh saw it, Panbanisha had a secret she urgently wanted to tell.

A decade and a half after the claims of animal language researchers were discredited as exaggerated self-delusions, Dr. Savage-Rumbaugh is reporting that her chimpanzees can demonstrate the rudimentary comprehension skills of two-and-a-half-year-old children. According to a series of papers, the bonobo, or pygmy, chimps, which some scientists believe are more humanlike and intelligent than the common chimpanzees studied in the earlier, flawed experiments, have learned to understand complex sentences and use symbolic language to communicate spontaneously with the outside world.

"She had never put those three lexigrams together," Dr. Savage-Rumbaugh said, referring to the keyboard symbols with which the animals are trained. She found the incident particularly gratifying because the chimp seemed to be using the symbols not to demand food, which is usually the case in these experiments, but to gossip.

In a book published by Routledge, *Apes, Language and the Human Mind: Philosophical Primatology,* Dr. Savage-Rumbaugh and her coauthors, Dr. Stuart G. Shanker, a philosopher at York University in Toronto, and Dr. Talbot J. Taylor, a linguist at the College of William and Mary in Virginia, argue that the feats of the chimps at the Language Research Center are so impressive that scientists must now reevaluate some of their most basic ideas about the nature of language.

Most language experts dismiss experiments like the ones with Panbanisha as exercises in wishful thinking. "In my mind this kind of research is more analogous to the bears in the Moscow circus who are trained to ride unicycles," said Dr. Steven Pinker, a cognitive scientist at the Massachusetts Institute of Technology who studies language acquisition in children. "You can train animals to do all kinds of amazing things." He is not convinced that the chimps have learned anything more sophisticated than how to press the right buttons in order to get the hairless apes on the other side of the console to cough up M&M's, bananas and other tidbits of food.

Dr. Noam Chomsky, the MIT linguist whose theory that language is innate and unique to people forms the infrastructure of the field, says that attempting to teach linguistic skills to animals is irrational—like trying to teach people to flap their arms and fly.

"Humans can fly about thirty feet—that's what they do in the Olympics," he said in an interview. "Is that flying? The question is totally meaningless. In fact the analogy to flying is misleading because when humans fly thirty feet, the organs they're using are kind of homologous to the ones that chickens and eagles use." Arms and wings, in other words, arise from the same branch of the evolutionary tree. "Whatever the chimps are doing is not even homologous as far as we know," he said. There is no evidence that the chimpanzee utterances emerge from anything like the "language organ" Dr. Chomsky believes resides only in human brains. This neural wiring is said to be the source of the universal grammar that unites all languages.

But some philosophers, like Dr. Shanker, complain that the linguists are applying a double standard: they dismiss skills—like putting together a noun and a verb to form a two-word sentence—that they consider nascent linguistic abilities in a very young child.

"The linguists kept upping their demands and Sue kept meeting the demands," said Dr. Shanker. "But the linguists keep moving the goal post."

Following Dr. Chomsky, most linguists argue that special neural circuitry needed for language evolved after man's ancestors split from those of the chimps millions of years ago. As evidence they note how quickly children, unlike chimpanzees, go from cobbling together two-word utterances to effortlessly spinning out complex sentences with phrases embedded within phrases like Russian dolls. But Dr. Shanker and his colleagues insist that Dr. Savage-Rumbaugh's experiments suggest that there is not an unbridgeable divide between humans and the rest of the animal kingdom, as orthodox linguists believe, but rather a gradation of linguistic skills.

In the book *The Engine of Reason, the Seat of the Soul: A Philosophical Journey Into the Brain* (MIT Press), Dr. Paul Churchland, a philosopher and cognitive scientist at the University of California at San Diego, says linguists should take Dr. Savage-Rumbaugh's experiments as a challenge. He argues that the jury is still out: the rules for constructing sentences might turn out to be not so much hardwired as a result of learning—by people and potentially by their chimpanzee relatives.

Animal language research fell into disrepute in the late 1970s when "talking" chimps like Washoe and the provocatively named Nim Chimpsky were exposed as unintentional frauds. Because chimpanzees lack the vocal apparatus to make a variety of modulated sounds, the animals were taught a vocabulary of hand signs—an approach first suggested in the 18th century by the French physician Julien Offray de La Mettrie. In appearances on television talk shows, trainers claimed the chimps could construct sentences of several words. But upon closer examination, scientists found strong evidence that the chimps had simply learned to please their teachers by contorting their hands into all kinds of configurations. And the trainers, straining to find examples of linguistic communication, thought they saw words among the wiggling, like children seeing pictures in the clouds.

In a widely quoted paper in the journal Science, "Can an Ape Create a Sentence?" Nim Chimpsky's trainer, Dr. Herbert Terrace, a Columbia University psychologist, reluctantly concluded that the answer was no.

A chimp might learn to connect a hand sign with an item of food, skeptics like Dr. Terrace argued, but this could be a matter of simple conditioning, like Pavlov's dogs learning to salivate at the sound of a bell. Most important, there was no evidence that the chimps had acquired a generative grammar—the ability to string words together into sentences of arbitrary length and complexity.

As a young veteran of the original animal language experiments, Dr. Savage-Rumbaugh decided to try a different approach. To eliminate the ambiguity of hand signs, she used a keyboard with dozens of buttons marked with geometric symbols.

In elaborate exercises beginning in the mid-1970s, she and her colleagues taught common chimpanzees and bonobos to associate symbols with a variety of things, people and places in and around the laboratory. The smartest chimps even seemed to learn abstract categories, identifying pictures of objects as either tools or food. Dr. Savage-Rumbaugh reported that two of the chimps learned to use symbols to communicate with each other. Pecking away at the keyboard, one would tell a companion where to find a key that would liberate a banana for them both to share.

Most impressive of all was a bonobo named Kanzi. After futilely trying to train Kanzi's adopted mother to use the keyboard, the researchers found that the two-and-a-half-year-old chimp, who apparently had been eavesdropping all along, had picked up an impressive vocabulary on his own. Kanzi was taught not in laboriously structured training sessions but on walks through the 50 acres of forest surrounding the language center. By the time he was six years old, Kanzi had acquired a vocabulary of 200 symbols and was constructing what might be taken as rudimentary sentences consisting of a word combined with a gesture or occasionally of two or three words. Dr. Savage-Rumbaugh became convinced that exposure to language must start early and that the lessons should be driven by the animal's curiosity.

Compared with other chimps, Kanzi's utterances are striking, but they are still far from human abilities. Kanzi is much better at responding to vocal commands like "Take off Sue's shoe." In one particularly arresting feat, recorded on videotape, Kanzi was told, "Give the dog a shot." The chimpanzee picked up a hypodermic syringe lying on the ground in front of him, pulled off the cap and injected a toy stuffed dog.

Dr. Savage-Rumbaugh's critics say there is nothing surprising about chimpanzees or even dogs and parrots associating vocal sounds with objects. Kanzi has been trained to associate the sound "dog" with the furry thing in front of him and has been programmed to carry out a stylized routine when he hears "shot." But does the chimp really understand what he is doing?

Dr. Savage-Rumbaugh insists that experiments using words in novel contexts show that the chimps are not just responding to sounds in a knee-jerk manner. It is true, she says, that Kanzi was initially aided by vocal inflections, hand gestures, facial expressions and other contextual clues. But once it had mastered a vocabulary, the bonobo could properly respond to 70 percent of unfamiliar sentences spoken by a trainer whose face was concealed.

None of this is very persuasive to linguists for whom the acid test of language is not comprehension but performance, the ability to use grammar to generate ever more complex sentences.

Dr. Terrace says Kanzi, like the disappointing Nim Chimpsky, is simply "going through a bag of tricks in order to get things." He is not impressed by comparisons to human children. "If a child did exactly what the best chimpanzee did, the child would be thought of as disturbed," Dr. Terrace said.

The scientists at the Language Research Center are "studying some very complicated cognitive processes in chimpanzees," Dr. Terrace said. "That says an awful lot about the evolution of intelligence. How do chimpanzees think without language, how do they remember without language? Those are much more important questions than trying to reproduce a few tidbits of language from a chimpanzee trying to get rewards."

Attempting to shift the fulcrum of the debate over performance versus comprehension, Dr. Savage-Rumbaugh argues that the linguists have things backward: "Comprehension is the route into language," she says. In her view it is easier to take an idea already in one's mind and translate it into a grammatical string of words than to decipher a sentence spoken by another whose intentions are unknown.

Dr. Shanker, the York University philosopher, believes that the linguists' objections reveal a naive view of how language works. When Kanzi gives the dog a shot, he might well be relying on all kinds of contextual clues and subtle gestures from the speaker, but that, Dr. Shanker argues, is what people do all the time.

Following the ideas of the philosopher Ludwig Wittgenstein, he argues that language is not just a matter of encoding and decoding strings of arbitrary symbols. It is a social act that is always embedded in a situation.

But trotting out Wittgenstein and his often obscure philosophy is a way of sending many linguists bolting for the exits. "If higher apes were incapable of anything beyond the trivialities that have been shown in these experiments, they would have been extinct millions of years ago," Dr. Chomsky said. "If you want to find out about an organism you study what it's good at. If you want to study humans you study language. If you want to study pigeons you study their homing instinct. Every biologist knows this. This research is just some kind of fanaticism."

There is a suspicion among some linguists and cognitive scientists that animal language experiments are motivated as much by ideological as scientific concerns—by the conviction that intelligent behavior is not hardwired but learnable, by the desire to knock people off their self-appointed thrones and champion the rights of downtrodden animals.

"I know what it's like," Dr. Terrace said. "I was once stung by the same bug. I really wanted to communicate with a chimpanzee and find out what the world looks like from a chimpanzee's point of view."

—GEORGE JOHNSON, June 1995

She Talks to Apes and, According to Her, They Talk Back

DR. EMILY SUE SAVAGE-RUMBAUGH, 52, a researcher at Georgia State University in Decatur, Georgia, studies communication among primates and runs a 55-acre laboratory near Atlanta where she trains animals and humans to communicate with each other.

She is the author of *Kanzi: The Ape at the Brink of the Human Mind,* and, with Stuart G. Shanker and Talbot J. Taylor, is a coauthor of *Ape Language and the Human Mind,* published by Oxford University Press.

Q. Do your apes speak?

A. They don't speak. They point to printed symbols on a keyboard. Their vocal tract isn't like ours, and they don't make human noises. However, they do make all kinds of ape noises. And I believe they use them to communicate with one another. Now, the apes may not always elect to talk about the same things we do. They might not have a translation for every word in our vocabulary to theirs. But from what I've seen, I believe they are communicating very complex things.

Let me give you an example. A few weeks ago, one of our researchers, Mary Chiepelo was out in the yard with Panbanisha. Mary thought she heard a squirrel and so she took the keyboard and said, "There's a squirrel." And Panbanisha said "DOG." Not very much later, three dogs appeared and headed in the direction of the building where Kanzi was.

Mary asked Panbanisha, "Does Kanzi see the dogs?" And Panbanisha looked at Mary and said, "A-frame." A-frame is a specific sector of the forest here that has an A-frame hut on it. Mary later went up to "A-frame" and found the fresh footprints of dogs everywhere at the site. Panbanisha knew where they were without seeing them.

And that seems to be the kind of information that apes transmit to each other: "There's a dangerous animal around. It's a dog and it's coming towards you."

Q. Your apes watch a great deal of TV—why?

A. Because their lives are so confined. They can expand their world by watching television.

Q. What do they watch?

A. This varies. They like the home videos we make about events happening to people they know from around the lab. They like suspenseful stories, with an interesting resolution. Of movies we buy, they really like films about human beings trying to relate to some kind of apelike creatures. So they like *Tarzan*, *Iceman*, *Quest for Fire*, the Clint Eastwood movies with the orangutan.

Q. You have a game with the apes, "Monster," where a lab staffer dresses up in a gorilla suit and feigns being frightful. Why?

A. It's a game started some years ago when we were working with two chimps, Sherman and Austin. We discovered that if someone dressed up in a gorilla suit and we drove this "monster" off with poundings of hammers and sticks, we upped our status with the chimps. In other words, "We're not the experimenters, in charge. We're your helpers." Sherman and Austin didn't know we were playing. For a while Kanzi and Panbanisha didn't either. But they caught on soon enough and now they love the game. . . .

. . . Another time, Panbanisha and I were walking around the building where Sherman and Mercury, this male chimpanzee with a big interest in Panbanisha, live. Mercury came outside and was being really bad, displaying, throwing bark, and spitting at Panbanisha. So Panbanisha opened her backpack, where there was a gorilla mask inside and she pointed to symbols on the keyboard and asked Mary to play "Monster." Mary did that, and Mercury flew indoors.

Panbanisha was able to use the game to stop him from displaying at her. She knew it was pretend. He didn't.

Q. How do you know when the chimps point to symbols on the keyboard that they are not just pointing to any old thing?

A. We test Kanzi and Panbanisha by either saying English words or showing them pictures. We know that they can find the symbol that corresponds to the word or the picture. If we give similar tests to their siblings who haven't learned language, they fail.

Many times, we can verify through actions. For instance, if Kanzi says "Apple chase," which means he wants to play a game of keep away with an apple, we say, "Yes, let's do." And then, he picks up an apple and runs away and smiles at us.

Q. Some of your critics say that all your apes do is mimic you.

A. If they were mimicking me, they would repeat just what I'm saying, and they don't. They answer my questions. We also have data that shows that only about two percent of their utterances are immediate imitations of ours.

Q. Nonetheless, many in the scientific community accuse you of over-interpreting what your apes do.

A. There are *some* who say that. But none of them have been willing to come spend some time here. I've tried to invite critics down here. None have taken me up on it. I've invited Tom Sebeok (of Indiana University) personally and he never responded. I think his attitude was something to the effect that, "It's so clear that what is happening is either cued, or in some way overinterpreted, that a visit is not necessary." I would assume that many of the people associated with the Chomskyian perspective including Noam Chomsky himself have the same approach: that there's no point in observing something they're certain doesn't exist.

Their belief is that there is thing called human language and that unless Kanzi does everything a human can, he doesn't have it. They refuse to consider what Kanzi does, which is comprehend, as language. And it's not even a matter of disagreeing over what Kanzi does. It's a matter of disagreeing over what to call these facts. They are asking Kanzi to do everything that humans do, which is specious. He'll never do that. It still doesn't negate what he can do.

Q. Your husband, Dr. Duane M. Rumbaugh, is a distinguished comparative psychologist who is a pioneer in the study of ape language. Has your research been helped by the fact that your personal life is so fused with your professional life?

A. Without our being together, I don't think that one could ever be responsible for as many apes as we have here. Duane and I live right near the research center and we're willing to go there day and night, 365 days a year. If an ape is sick, if one of the apes has gotten free, if Panbanisha is frightened because she's heard the river's about to flood, we go.

There have been lots of frictions, though. Duane was very, very upset when I began taking the apes out of their cages. And when I began to say that Lana (Duane's chimp) didn't understand some of the things she was saying and that comprehension of language was important, not just production—we almost broke up over that.

But we really love each other, and we're united in our core beliefs: that there is a huge capacity on the part of apes and probably all kinds of other animals that's being ignored. By ignoring it, humans are separating ourselves from the natural world we've evolved from. The bonobos are a real bridge to that world. At base, no matter how much Duane and I argue, we both know this is true.

—CLAUDIA DREIFUS, April 1998

Picking Up Mammals' Deep Notes

WATCHING ELEPHANTS FROM A TOWER in the African savanna seems a long way from the music studies Katy Payne pursued here as an undergraduate student at Cornell University in the 1950s. But she insists that her career as a field biologist, which has produced remarkable discoveries about the language of the world's largest mammals, whales and elephants, has followed "a natural, logical line," and that her knowledge of music has been an essential ingredient in her studies of animal communication.

"It's all connected—my outdoorsy childhood, my early musical experiences, my marriage to a musical biologist who abandoned laboratory science to study whales," she said. Even her divorce in 1985 from the biologist, Dr. Roger Payne, fostered new discoveries by prompting Ms. Payne, a self-taught bioacoustician, to turn her research attention from whales to elephants.

Now Ms. Payne is hoping her work with elephants will help save these majestic beasts from continued destruction. The summer of 1986 found her perched above a semidesert in Namibia, her fine-tuned senses alert to the elephants at the water hole below. From a loudspeaker attached to a van in Etosha Park more than a mile away came a single 40-second broadcast not audible to human ears. It was a sequence of low-frequency calls of a female elephant in heat, an event that happens only once every four or five years and lasts for only two to four days.

Two bull elephants at the water hole suddenly stopped drinking, raised their heads, perked up their ears and turned in the direction of the van. In a moment they began walking—fast, for elephants—toward the van. Though the broadcast was not repeated, the bulls marched on so resolutely that the scientists in the van began to worry whether they would be safe when the lustful bulls arrived.

Fortunately, the disappointed bulls walked right past the truck. The research team started breathing again and rejoiced in what the experiment had shown: African elephants respond over long distances, at least two and a half miles, according to the scientists' extrapolation, to one another's infrasonic calls, very low-frequency sounds below the range of human hearing but excellent for long-distance transmission through forests and grasslands.

Ms. Payne's investigations of elephant sounds unveiled an unusual and possibly unique form of communication among mammals. If not for her extraordinary sensitivity and ability to relate disparate elements, she might never have "heard" the silent sounds that form the basis for the elephant language she is elucidating.

Born Katharine Boynton on a farm here in 1937 to a father who was an apple grower and Cornell professor and a mother who loved books and music, Katy Payne was a sickly child who missed a lot of school and instead spent much of her youth exploring the gorges of Ithaca and learning to appreciate the creatures painted by her grandfather, the famous wildlife illustrator Louis Agassiz Fuertes.

As a student at Cornell, although intensely interested in nature, she majored in music, she said, "because I was turned off by what was then the molecular orientation of biology." Then she met Roger Payne, a graduate student in biology who was playing the cello at a concert—a performance of the "St. Matthew Passion"—in which she was singing. For Ms. Payne, it was the first of a series of serendipitous experiences that have helped make her preeminent in the field of animal communication, despite a lack of formal training and advanced degrees.

Marriage and the arrival of four children in four years ended her attempt to earn a graduate degree in biology. But when her husband's whale song research got under way, she spent hours each day analyzing the tapes. She and Roger Payne began working together in the field, taking the children out of school for extended research trips to Bermuda, Hawaii and eventually Patagonia, where the Paynes launched studies of the Southern right whale.

"It didn't seem to hurt them academically," Ms. Payne says of her children's extended absences from school. Indeed, she adds proudly, "all four of my children are in careers that involve wildlife conservation."

The Paynes had discovered the songs of humpback whales during a brief trip in 1968 to Bermuda, where Frank Watlington, a Navy engineer who had been monitoring an array of hydrophones many miles into the sea, decided to trust them with his best-kept secret. While the three watched a pair of humpbacks disappear into the warm sea, he confided, "You know they make sounds, don't you?" He then took them to the bowels of his ship and played some of his tapes.

Mr. Watlington had feared that whalers would use the whale sounds to help locate their prey and so had told only a few trusted friends about the hypnotic songs his hydrophones had picked up. As it turned out, through the work of the Paynes and their collaborators, the whale songs, which were captured on a bestselling nature recording, aroused intense public interest in and concern about whales, which helped curtail the whaling industry long enough to allow the beginning of a recovery.

Of Ms. Payne's contribution to the work, Dr. Peter Marler, professor of neurobiology and behavior at the University of California at Davis, said, "She really was the one who worked out the most extraordinary findings about whale songs: the fact that they rhyme and the predictable ways in which the whales change their songs each season." He said she had the patience and tenacity to pore over 31 years of recordings, making cutouts of the musical lines spewed out by a spectrograph, painstakingly analyzing and overlapping them, and listening over and over again until she could discern the individual voices in a whale quartet and then write out the score.

Her findings demonstrated that although at any one time all the whales in the same large area of the ocean sing the same song and all change their song in the same way, they do not all sing the same number of phrases. "It's like a round in which some of the singers leave out some of the lines," she explained. Yet amid the seeming cacophony there is remarkable order, with each whale singing the same themes in the same sequence, each changing the phrases in the same way as the singing season progresses and each resuming the next season with roughly the same song he had sung when the previous season ended.

With Linda Guinee, then an independent researcher, Ms. Payne also found that the more involved the songs were, the more likely they were to have repeating refrains that constituted rhymes. This finding suggested that the whales were using rhyme the way people do: as a mnemonic device to help them remember complex material.

Studies by Dr. Peter Tyack, an associate scientist at the Woods Hole Oceanographic Institute in Woods Hole, Massachusetts, are beginning to reveal what the songs, which are sung only by males, mean to the lives of whales. They seem to enable males to establish floating territories and may function somehow in mating, although no one has yet seen humpback whales mate. Still, the Paynes' research has clearly demonstrated that all the males in a given part of the ocean listen carefully to each other and follow the same grammatical rules as their language evolves.

One fall day, Ms. Payne met with a visitor in her lab in a big old barn at Cornell's Laboratory of Ornithology, which runs the university's Bioacoustics Research Program. Before joining Cornell as a research associate in 1984, and after her collaborative work with Roger Payne ended, she was casting about for a new field of research. She made a visit to the Metro Washington Zoo in Portland, Oregon, where she spent a week in the elephant enclosure watching and tape-recording three mothers with their newborn calves. The week was disappointing and she came away thinking she had seen and heard little of research interest.

Then, during the plane ride home, she suddenly realized that though she had heard little, she had felt a lot—a lot of vibrations, as if from distant thunder or an earthquake. The sensations were familiar to her, from very long before, when as a teenager she and her mother had sung the "St. Matthew Passion" in a chapel choir. She had stood near the largest pipes of the organ and had felt throbbing vibrations fill the chapel when the organist played the lowest notes. Maybe, she thought, the elephants were speaking in frequencies even lower than the organ's, frequencies that people could feel but not hear.

Soon after her visit to Portland, with the support of the World Wildlife Fund and the Cornell University Laboratory of Ornithology, she returned to the Portland zoo with two old friends, Dr. William R. Langbauer, then a biologist in New York, and Elizabeth Thomas, an anthropologist in Peterborough, New Hampshire, and author of, among other books, the bestseller *The Hidden Life of Dogs* (Davison/Houghton Mifflin). This time the researchers were equipped with a tape recorder that picked up infrasound. When the scientists later studied an electronic printout of their tapes, they discovered an amazing array of sounds, many times more than they had been able to hear. The elephants were the first land mammals known to use low-frequency sounds.

Ms. Payne was off and running. With an invitation from Dr. Joyce Poole and Cynthia Moss to join their behavioral studies of elephant families in Amboseli National Park in Kenya and later with grants from the National Geographic Society, World Wildlife Fund and the National Science Foundation, she, Dr. Langbauer and Russ Charif of Cornell's Bioacoustics Research Program began to tackle elephant behavior in a new way. Ms. Payne and her colleagues wanted to know what elephants say to one another, when and why, and over what distances.

There are many mysteries about elephant behavior. The females live in independent family groups, often miles from the males. Yet Dr. Poole had observed that very soon after going into heat, females are surrounded by males. How do they find one another during those few precious days of female receptivity? Another mystery is how distant matriarchal groups coordinate their movements. Rowan Martin, a government employee in Zimbabwe, had tracked radio-collared elephants for years and found that though separated by miles, elephant families seemed to move in the same direction for days at a time.

By recording more than 1,000 calls emitted by Amboseli elephants, most at frequencies of 14 to 35 hertz (human hearing starts at about 30 hertz), and by coordinating these infrasounds with observations of the animals' behavior, Ms. Payne and Dr. Poole identified a greeting rumble, contact call and answer, a "let's go" rumble, a rumble uttered by males during periods of heightened sexuality, a female chorus that replies to this rumble and a song sung by females in heat, which may be repeated like a lovesick refrain for up to 40 minutes.

In another study in Zimbabwe, Ms. Payne and Dr. Langbauer fitted 16 female elephants from 13 different families with radio collars that had voice-activated microphones to enable the sounds each elephant made to be associated with her movements. The data are still being analyzed, but thus far they suggest that the animals sometimes coordinate their movements over long distances, possibly using infrasound.

"Basically, I think elephants listen passively to each other, using sounds to space themselves so that all get enough to eat but at the same time keeping the bond group within earshot so that they can respond quickly to danger," Ms. Payne said.

Ms. Payne has written a children's science book, *Elephants Calling* (Crown), illustrated with her own photographs, and she is writing a book for adults about the lives of elephants and the roots of conservation in

Africa. Grants permitting, her future will include studies of communication among forest elephants in the Central African Republic and Asian elephants, both wild and domesticated.

—JANE E. BRODY, November 1993

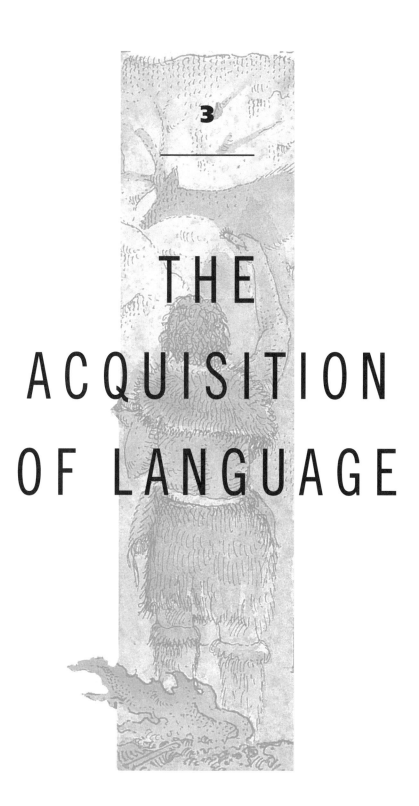

3

THE ACQUISITION OF LANGUAGE

If humans alone have the capacity for language, what is it in the brain that confers the power of speech?

From studying babies, biologists believe that infants acquire language in an orderly pattern, as if some fixed program of neural development were being played out. If language is not learned during a critical period of time, the capacity for learning it may be lost, as happens occasionally with children who have been isolated or ignored as infants.

The rules of syntax may be wired into the brain, but each language is learned individually. Babies seem attentive from their earliest months to the sound of the language spoken around them, even though they do not comprehend it. The stylized way in which adults talk to infants seems designed to reinforce some pattern that is important for the babies' later speech abilities. Even deaf babies, it turns out, make repetitive motions with their hands that seem to be the manual equivalent of babbling.

But despite many intriguing insights into the early acquisition of language, biologists are a long way from being able to offer scientifically based advice about child rearing.

In Brain's Early Growth, Timetable May Be Crucial

FOR THE FIRST 28 MONTHS OF her life, Simona Young languished in a Romanian orphanage. She lay in a crib alone for up to 20 hours a day, sucking nourishment from cold bottles propped over her tiny body. Unable to sit up by herself, she would push her torso up on thin arms and rock back and forth for hours, trying to soothe the aching void that had replaced her mother.

Now six, she runs, talks and sings like other children her age. Since she was adopted by a Canadian family in 1991, she has been making steady progress, says her new mother, Jennifer Young. Yet problems remain. Simona still suffers temper tantrums and has trouble following spoken directions. She has difficulty sharing and taking turns with other children, and she will happily wander off with strangers who say kind words to her.

Psychologists at Simon Fraser University in Burnaby, British Columbia, are closely watching Simona's development, along with that of 44 other Romanian orphans who were adopted around the same time. Like Simona, 30 of the children experienced one or more years of profound deprivation in the overcrowded orphanage, where staff workers gave infants little or no personal attention. The other 15 were adopted within a month or two of their births.

The Canadian researchers are comparing the two groups of children to help answer an age-old question: can love overcome a bad beginning?

Other scientists are asking similar questions, using the tools of modern cognitive neuroscience: are there very early critical periods for emotional development? How does experience shape the brain's circuits? How changeable are those circuits later in life?

No one is saying there are quick fixes, that making nice to a baby between birth and 24 months will avert all later problems, said Dr. Carla Shatz, a developmental biologist at the University of California at Berkeley and president of the Society for Neuroscience. But basic brain research is seeking answers that may ultimately help guide social policy, she said.

If there are critical periods for a child's emotional development, parents may be taught when and how to provide the kind of nurturing needed for healthy brain development. If the adult brain is amenable to change, maladaptive circuits formed in infancy or childhood may be alterable by psychotherapy or other methods.

Much is already known. Even the human fetus can hear sounds and has limited vision, Dr. Shatz said. "The nervous system isn't waiting for birth to flip a switch and get going," she said.

In the 1960s, Dr. David Hubel and Dr. Torsten Wiesel found that vision does not develop normally in cats if the eye and brain fail to make connections during a critical window of time in early life. Kittens that had one eye kept closed after birth did not develop the usual connections between that eye and the primary visual area of the brain. Once this period, lasting several weeks, had passed, none of the kittens could see out of the eye that had been closed, even though it was perfectly normal.

Hearing and language are also abilities that develop during critical periods, Dr. Shatz said. A Japanese baby can distinguish "r" from "l," but, absent the "l" sound in the Japanese language, loses this ability after age three. After 10, most people cannot learn to speak a second language without an accent. Unless deaf children are exposed to some form of language before age five, they behave as though they are retarded. And so-called "wild" children, raised without human contact, never learn to speak with fluency.

In recent years, the search for critical or sensitive windows of development has extended to other biological systems in the brain.

All animals, including humans, develop a control point in early infancy for how much of various stress hormones they will release in particular conditions, said Dr. Michael Meaney, a psychiatrist at McGill University in Montreal. Animals experiencing high stress levels in infancy develop a highly reactive system, he said, while animals raised in relative calm have quieter systems.

Dr. Myron Hofer, a psychiatrist at the New York State Psychiatric Institute in New York, has found numerous "hidden modulators" in the mother-infant relationship. For example, the licking of a mother rat influences the setting of her pup's heart rate, temperature, circadian rhythms, growth, immune system and other physiological states.

Other researchers are studying how a mother's touch literally helps shape her baby's brain. If baby rats are deprived of maternal licking when they are seven to 14 days old, they develop fewer hormone receptors in their brains. Missing the needed stimulation in this critical period, they fail to grow normally, even when adequate amounts of growth hormone and insulin circulate in their tissues.

Human mothers provide similar modulators, Dr. Hofer said, through rocking, touching, holding, feeding and gazing at their babies. Some of these regulators are emotional, he said; thus, a baby knows when its mother is being cold or distant, despite her ministrations to physical needs. In the first six months of life, "the infant is laying down a mental representation of its relationship with its mother," Dr. Hofer said, adding, "These interactions regulate the infant's neural mechanisms for behavior and for feelings that are just beginning to develop."

If these early months of life are so important, what is actually happening inside the baby's brain? What kinds of changes are taking place?

At birth, according to Dr. Harry Chugani then of the University of Michigan in Ann Arbor, a newborn brain has fewer synapses—connections between nerve cells—than an adult brain. (The same holds true for the complexity of dendrites, or branches.) But the number of synapses reaches adult levels by age two and continues to increase, far surpassing the adult level from ages four to 10, Dr. Chugani said. The density of synapses then begins to drop, returning to typically adult levels by age 16. These findings are based on direct anatomical measurements by Dr. Peter Huttenlocher of the University of Chicago, who measured the brains of children killed in car accidents, and on brain images from PET scans that Dr. Chugani performed on infants for health reasons.

Concurrent with the explosion in the growth of synapses is a rapid pruning away of those that do not get used, Dr. Chugani said. There seems to be as much synaptic death as there is synaptic profusion.

The interplay between genes and experience in building a complex structure like the brain is to be expected, said Dr. Daniel Alkon, chief of the Neural Systems Laboratory at the National Institutes of Health in Bethesda, Maryland. Human DNA does not contain enough information to specify how the brain finally gets wired.

Thus the newborn brain comes equipped with a set of genetically based rules for how learning takes place and is then literally shaped by experience, Dr. Alkon said. "This helps explain the power of childhood memories," he said. "Associations in early life help choose which synapses live or die."

Dr. Jeff Shrager, a neuroscientist at the University of Pittsburgh, says the infant's brain seems to organize itself under the influence of waves of so-called trophic factors—chemicals that promote the growth and interconnections of nerve cells. These factors are released so that different regions of the brain become connected sequentially, with one layer of tissue maturing before another and so on until the whole brain is mature. Such waves of chemical activity may help determine the timing of critical periods.

By the time the brain's production of trophic factors declines in later childhood, its basic architecture would be more or less formed, Dr. Shrager said. The process, since modulated by experience, would create human brains that were similar in their overall structure and interconnections but unique in terms of their fine connections.

The same trophic factor chemistry that makes young brains grow so dramatically may still be available in adulthood, particularly in the hippocampus, to help with adult learning and memory, Dr. Shatz said. It is thus possible that brain circuits carved by early experiences may be changed through psychotherapy or other means.

While that hope remains, there is a deeper question yet to be answered: are there narrow windows in early infancy when emotional circuits are permanently established, or do emotional circuits form over many years so that early experiences are not so powerfully formative?

Much of the thinking is still speculative, Dr. Alkon said. "But we do know that a child learns trust and self-worth in the first two years," he said. "When a parent neglects a baby on a daily basis, the child is conditioned to expect isolation. A recipe for depression has been acquired from experience, handed down from one generation to another."

Pet scans show that the frontal cortex becomes metabolically very active in infants aged six to 24 months, Dr. Chugani said, and again at puberty.

Thus it is possible that the frontal cortex—once thought to develop in later childhood—may be involved in early emotional and cognitive development, said Dr. Geraldine Dawson, a psychologist at the University of Washington in Seattle.

Dr. Dawson and others have found that the left frontal lobe is activated when a person feels happiness, joy or interest, while the right is associated with negative feelings. Infants of severely depressed mothers show reduced activity in the left frontal region, she said. Activity in the right is increased, which means the babies are vulnerable to negative emotions.

"Our hunch is that there may be a critical period for emotional development between ages eight and 18 months," Dr. Dawson said. "This is when kids learn to regulate emotions. It is when attachment forms."

The insights into brain development gained from animal experiments might apply to humans, but in many cases repeating the experiments in children would be unethical. The Romanian orphans are of particular interest to brain researchers. The deprivations inflicted on them by the Romanian regime and, among the adopted children, the efforts of their new parents to nurture them back to normality, in effect constitute a unique experiment.

The children still in Romanian orphanages "look frighteningly like Harlow's monkeys," said Dr. Mary Carlson, a neuroscientist at Harvard University, referring to a well-known experiment of the 1950s in which baby monkeys were removed from their mothers a few hours after birth and reared without parental care. The infants developed abnormal behaviors. They often sat and stared for long periods, or would rock back and forth. Despite later efforts to rehabilitate them, the monkeys had disturbances in social behavior.

Many institutionalized Romanian orphans are below the third percentile in weight and height, Dr. Carlson said. Some show a profound failure to thrive, and at age 10 are the size of three-year-olds, suggesting that the absence of early maternal interaction has had lasting effects.

But humans being more adaptable than monkeys, researchers are striving to reverse the effects of deprivation in the orphans. Elinor Ames, a psychologist at Simon Fraser University, notes that the older children, now four and a half to 10 years old, are catching up in language and physical development. But they are having trouble with social development. When they are in stressful situations, she said, they wrap their arms around themselves and rock for comfort. Although most feel close to their adopted parents, their attachments are not always secure and some of the children will wander off with strangers.

The hope is that continuing good experiences will help these children grow into secure adults, Dr. Ames said, adding that she is optimistic.

But if it turns out that positive early experiences are crucial for healthy brain development and that deprivation leads to depression, anger and pathological behavior, what can society do to intervene? The question applies to babies being raised in violent neighborhoods by drug-addicted mothers as well as to poor little rich kids whose parents are too busy to pay them attention.

Intervention experiments—in which disadvantaged children are taken to special day care centers from infancy to kindergarten five days a week and given a rich educational curriculum and loving environment—have worked, said Dr. Craig Ramey, a psychologist and educator at the University of Alabama in Birmingham.

Effects of the intervention did not begin to show up until the second year of life, he said, but at age two a matched group of children who were not given the intervention had I.Q. scores 15 points below those who were helped.

"The results are clear," Dr. Ramey said. "To make a difference, you have to intervene earlier" than Head Start. "We think we are affecting early mechanisms involved in language acquisition, in the depth and breadth of the language experience which lays a foundation for higher order thinking" later in life, he said.

Dr. Carlson, whose work with Romanian orphans has led her into advocacy for children's rights, said that she had been advised by people who worked for government child welfare agencies to play down the notion of critical periods and early brain development. "They say, 'If so much is determined by age one or two, people will give up on children,'" she said. "But I think that if you believe in critical periods, you can find ways to take advantage of that plasticity."

Dr. Megan Gunnar, a professor of child development at the University of Minnesota in St. Paul, said the problem was not as difficult as people might think. "The one thing we have learned," she said, "is that children need to feel safe and protected. That alone leads to appropriate biological growth."

Dr. Hofer agreed. "If you grow up in battle-torn Yugoslavia, you may become impulsive, aggressive and you won't want to get close to anyone," he said. "You will be beautifully suited to fight a five-hundred-year war in Europe."

—SANDRA BLAKESLEE, August 1995

Linguists Debate Study Classifying Language as Innate Human Skill

A STUDY OF A DEAF child's linguistic abilities is stirring up an ancient debate over the nature of language. Is the human brain uniquely programmed to make and learn languages or does it simply pick up on ordered structures perceived when a child is first exposed to speech?

The subject, a nine-year-old boy named Simon, is uniquely appropriate for the experiment of asking whether language is learned or innate because he learned an error-riddled form of American Sign Language from his parents, who are also deaf, and a quite different sign language, with different grammatical rules, at his school. Despite the faulty teaching of American Sign Language, Simon signed the language with correct grammar, which the researchers see as evidence that he was drawing upon innate language ability.

The researchers studied Simon from the time he was two and a half years old to the time he was nine. They reported that he had signed in American Sign Language with the correct grammar, even though he had learned incorrect grammar from his parents.

To the obvious objection that Simon may have seen other people signing correctly in American Sign Language, the researchers reply that his parents were the only people whom he had seen signing in American Sign Language, apart from his parents' friends, who also signed incorrectly. His parents and their friends learned American Sign Language only as teenagers, an age at which languages are often learned inaccurately.

The investigators, Dr. Elissa L. Newport of the University of Rochester and Dr. Jenny L. Singleton of the University of Illinois, believe that Simon recognized complex patterns in the language on the basis of his parents' inconsistent use of the patterns. And, they say, Simon learned to use

Linguists Debate Study Classifying Language as Innate Human Skill

some complicated rules in ways that had eluded his parents. The research was presented at a meeting of the American Psychological Society in San Diego.

Dr. Newport said that the way Simon had deduced grammatical rules showed "exactly the kinds of things you would predict" from theories of how children develop language. But it has been very difficult to find evidence that these theories are correct.

Other investigators said they were intrigued by this case. Dr. Ursula Bellugi, a neuroscientist at the Salk Institute in La Jolla, California, said, "It has been hard to get really solid evidence of whether the brain is disposed in particular ways for learning languages." The story of Simon, she said, "is really exciting" because it is so scientifically clean. "I think it's very convincing," she added.

And Dr. Susan Goldin-Meadow of the University of Chicago said, "I think their data are incontrovertible."

But Dr. Jean Berko Gleason of Boston University, who is the editor of the standard linguistics textbook *The Development of Language,* said that she would not read so much into the case history. Simon, she said, "seemed to pick up on the regularities of the language," but he did not invent language structure out of whole cloth. And the study is based on just a single child, she added. "It's always interesting even if one child does something, but you never know if he's showing universal tendencies," Dr. Berko Gleason said.

Simon's story is part of a centuries-long tradition of case studies of children who scientists hoped could help shed light on the question of whether language is innate and whether there is only a window of time, when children are maturing, in which it can be learned. Researchers have studied feral children, who are called that because they grew up with only animals for company. They have studied abused children who had been kept isolated and deprived of human talk and companionship. They have studied deaf children who had not been taught to sign.

But these studies were not scientifically pure, researchers said. The feral children and abused children had so many other emotional and physical problems that it was impossible to say what was cause and what was effect. The deaf children developed a language so simple that some question whether it counts.

Simon, on the other hand, was loved and cherished and was taught a language by his parents. The only thing missing was consistently correct complex grammar and sentence structure.

Although Simon's parents were each born deaf, to hearing parents, they did not learn to sign as children. Instead, they were sent to schools that tried to teach them to read lips. Like most deaf people, they never succeeded in this endeavor. Only as teenagers did they learn American Sign Language, so they learned it imperfectly, which is common when people learn a new language at that age.

But, the researchers said, Simon divined grammatical rules that his parents could not grasp. The parents, for example, had trouble with verbs of motion, which they used correctly just 65 percent of the time; Simon signed the verbs correctly 90 percent of the time.

Simon's parents had no grasp of some grammatical rules of American Sign Language, like a rule called "topicalization," which allows the signer to make a word the topic of a sentence even though it is not the subject. With topicalization, Dr. Newport said, "You can say, 'John hit Mary,' but if you want to talk about Mary, you would say, 'Mary, John hit.' " While moving the word "Mary" to the front of the phrase, the signer makes a special facial expression, with lifted eyebrows and chin and drawn-up muscles under the nose.

Although Simon's parents never moved words in sentences, they would emphasize the first word in a sentence by making the special facial expression. "Simon's parents don't seem to know the movement rule," Dr. Newport said. But Simon, on the other hand, "does it perfectly correctly." Since Simon never saw the rule used correctly, "I don't think he's extracting it from the pattern of input. I think it's something he's born with," she added.

"There is a very rich argument that kids must somehow be equipped with a lot of biases that make them organize languages in particular ways," Dr. Newport said.

For example, Dr. Noam Chomsky of the Massachusetts Institute of Technology, who initiated the modern era in linguistics in the late 1950s with his studies of language structure, has argued that all the world's languages share common features that reflect a biological determinism. He believes that all children are surrounded by errors and incompleteness when learning language but that they pick out the rich grammatical structures, developing a grasp of language that goes beyond their exposure. "What they know is so far beyond what they've heard that they obviously created it themselves," he said.

In other studies, Dr. Goldin-Meadow and her colleagues examined the deaf children of hearing parents who had not been taught any sign language and asked whether they had made up a language for themselves. About 20 years ago, it was common for deaf children to be discouraged from learning to sign, Dr. Goldin-Meadow said. "The parents don't know sign language, and they want their child to be part of the hearing world. If the child learns sign language, they will eventually leave that world," she said. So the parents and schools tried to teach the children to read lips and to speak. "Very, very, very few children ever succeeded, but there was always that hope," Dr. Goldin-Meadow said.

But, in the meantime, the children would make up their own signs. The language was simple, involving pointing and gesturing. But, Dr. Goldin-Meadow found, the children would start to string gestures together, while the parents would only rarely elaborate on the signs. A child, for example, "would point at a cup and then make gestures for drinking. But a parent would only rarely combine gestures, and even if they did combine them they would not combine them with such order," Dr. Goldin-Meadow said.

Still other studies, by Dr. Bellugi, involve deaf people whose brains had been injured by strokes. She discovered that the same area of the left brain is used for all language, whether spoken or signed, even though sign language is such a spatial language. One woman studied by Dr. Bellugi had suffered a stroke that had injured the right hemisphere of her brain.

"She was an artist and she was unable to draw; she couldn't do perspective," Dr. Bellugi said. "But her signing was perfect." Dr. Bellugi said she was convinced that "language acquisition is very much biologically determined."

But Dr. Berko Gleason said that these studies, while intriguing, still left open the question of whether the capacity to learn language is innate. Although the deaf children invent a sign language when they are not taught to sign, she said, "whether that is language as we know it is open to question." And the brain-injury studies do not prove that the language area in the brain was there before people learned language. With virtually all the studies, Dr. Berko Gleason said, "it is always the same problem: there are very small samples, and things are very much in the eye of the beholder."

But Dr. Chomsky disagrees. He said the idea that the human brain is organized to make the learning of language innate "is strongly established." Although he did not need Simon to convince him, he said, Simon's case shows that children extract more from language than they are ever explicitly shown.

—GINA KOLATA, September 1992

Babies Learn Sounds of Language by Six Months

BABIES LEARN THE BASIC SOUNDS of their native language by the age of six months, long before they utter their first words, and earlier than researchers had thought, a study suggests.

The findings indicate that recognition of these sounds is the first step in the comprehension of spoken language. As a result, the researchers suggest, babies whose hearing is damaged by chronic ear infections may have lifelong language problems, and the way parents speak to their infants exerts important influences on language learning.

Previous studies suggested that infants' sound perception changes by about one year old, when children begin to understand that sounds convey word meanings.

The research, reported in the journal *Science,* was conducted by Dr. Patricia Kuhl of the University of Washington in Seattle and colleagues at Stockholm University in Sweden, the Massachusetts Institute of Technology and the University of Texas in Austin.

Newborns are language universalists, Dr. Kuhl said. Able to learn any sound in any language, they can distinguish all the sounds that humans utter. But adults are language specialists, she said. Exposure to their native language reduces their ability to perceive speech sounds that are not in that native tongue. Thus Japanese infants can hear the difference between the English sounds "la" and "ra," but Japanese adults cannot because their language does not contrast those sounds.

Dr. Kuhl said she and her colleagues set out to discover when, during language development, experience alters sound perception and to explore the nature of the change. She said she had thought it could be earlier than other researchers believed.

To test her idea, she used the concept of phonetic prototypes—idealized mental representations of the key sounds in a given language. An English prototype sound is the vowel linguists write as "i," pronounced as in the word "fee." When an adult English speaker hears something very close to this "i" sound (as when the sound is spoken by someone with a head cold), Dr. Kuhl said, the listener will hear the prototype "i" and not the slight variation. The prototype sound acts like a magnet, she said, pulling all similar sounds into one mental slot for language processing.

But the same is not true of foreign languages. Because English speakers have not memorized the prototype for a foreign vowel—like the Swedish vowel "y" (an EE-sound pronounced with front-rounded lips), they can discern when the vowel is pronounced slightly differently. They have no "magnet" that makes the sounds identical.

Using identical computer equipment to generate prototype Swedish and English sounds, Dr. Kuhl and her colleagues tested the magnet effect on 64 six-month-old babies in Sweden and the United States. During the experiment, each baby sat on its mother's lap and listened to pairs of "i" and "y" sounds. Babies were trained to look over their left shoulders when they heard a difference in the sounds (they would see a cute puppet bang a drum) and to ignore any sound pairs that seemed the same.

American babies routinely ignored the different pronunciations of "i" because they heard it as the same sound, Dr. Kuhl said. But they could distinguish slight variations in the "y" sounds.

The exact opposite was true of the Swedish babies, she said. They ignored the variations in "y" because they sounded the same, while they noticed the variations in "i."

The experiment confirms that linguistic experience in the first half year of life alters an infant's perception of speech sounds, Dr. Kuhl said. Infants show a significantly stronger magnet effect for their native language prototypes.

The study shows that phonetic perception does not depend on the emerging use of words, Dr. Kuhl said, and that language experience shapes perception far earlier than anyone expected.

The research calls attention to the language-tutoring role of parents, Dr. Kuhl said. By talking "motherese" with its high pitch, exaggerated intonation and clear pronunciation, she said, parents help babies acquire phonetic prototypes that are building blocks to language.

The study also underscores the importance of treating chronic ear infections in infants, Dr. Kuhl said. There is evidence that such infections may impair language development later in life.

—SANDRA BLAKESLEE, February 1992

Study Finds Baby Talk Means More Than a Coo

WHEN MOTHERS IN THE UNITED STATES, Sweden and Russia nuzzle their infants and croon baby talk, they are giving their babies a lot more than tender loving care, scientists have found. The women are producing highly exaggerated speech sounds that provide the basis for all subsequent learning of that language.

Baby talk is much more important than people realize, said Dr. Patricia Kuhl, chairwoman of speech and hearing sciences at the University of Washington in Seattle and the lead author of a report on the university of "parentese"—the singsong, exaggerated speech pattern that people use when talking to infants.

"Parentese has a melody to it," Dr. Kuhl said. "And inside this melody is a tutorial for the baby which contains exceptionally well-formed versions of the building blocks of language."

In a study published in the journal *Science,* Dr. Kuhl and her colleagues found that mothers in the three countries stretched out or exaggerated the pronunciation of three primary vowels when speaking to their infants—"ee," "ah" and "oo"—which are common to every spoken language in the world. This vowel stretching tunes the infant's brain to the widest possible range of vowel sounds, Dr. Kuhl said. As the baby learns the vowels of its mother tongue, the ability to hear subtle vowel distinctions in other languages is gradually lost.

Dr. Richard Aslin, a professor of brain and cognitive sciences at the University of Rochester, who was not involved in the research, said, "This is an important study showing that what mothers do naturally can help babies. But it does not yet prove that the exaggerated input causes language to develop. You'd have to show that infants who do not hear parentese do not acquire language as well or as swiftly."

Dr. Aslin said that such a study would be difficult since parents in every culture speak some form of baby talk to their infants.

Five years ago, Dr. Kuhl and her colleagues reported that by the age of six months, babies had learned the specific sounds of their mother tongue, suggesting that cells in the brain region that processes sounds are tuned to the frequency of vowels and consonants in that language. Parentese seemed to literally alter the baby's brain tissue.

How this happened remained a question.

"We knew that pitch and melody are altered in parentese," Dr. Kuhl said in a telephone interview. The question was, she said, "are the phonetic units—the sounds of language—modified in any way so as to enhance learning?"

In other words, she said, "Is there more than melody to the message?"

To explore this question, the researchers used the concept of vowel space, a standard conceptual framework for studying the sounds of a language. All languages contain three so-called point vowels—"ee" as in "bead," "ah" as in "pot" and "oo" as in "boot." Although these vowels differ ever so slightly in pronunciation from one language to the next, they tend to be pure sounds that are extremely similar.

In the concept, the three vowels are plotted as points on a triangle, Dr. Kuhl explained, and the sounds of all other vowels fall within the triangle. Where they fall depends on their frequencies—essentially complex mixtures of "ees," "ahs" and "oos" that comprise the specific sound system of a language. The vowels of every language fall within the triangle. Swedish has 16 vowels, English has nine and Russian has five.

In the study, Dr. Kuhl and her colleagues recruited 30 mothers, 10 from each country, and audiotaped them speaking first to their infants, two to five months old, and then an adult.

Whether the mothers spoke English, Swedish or Russian, all greatly exaggerated their "oos," "ahs" and "ees" when speaking to their babies, Dr. Kuhl said. As measured by vowel frequency, the mothers doubled the size of the triangle in speaking to their babies as compared to speaking with adults. The mothers did not produce identical super vowels, but they exaggerated the three vowels by the same degree.

These exaggerated vowels are important for several reasons, Dr. Kuhl said. First, it makes it easier for babies to hear distinctions in speech. Sounds are more easily contrasted and this benefits the brain mapping of language. Second, once an infant's brain is tuned to super vowels, it is easier for the infant to hear the less distinct vowels of normal speech. Third, when mothers produce these sounds, they are giving their babies more "room" for hearing more sounds.

"It's like looking at a caricature of a face rather than at a real face," Dr. Kuhl said. "The caricature emphasizes the critical features of the face by exaggerating them and the same is true for sounds of a language."

—SANDRA BLAKESLEE, August 1997

Studies Show Talking with Infants Shapes Basis of Ability to Think

WHEN a White House conference on early child development convened, one of the findings Hillary Rodham Clinton heard from scientists is that the neurological foundations for rational thinking, problem solving and general reasoning appear to be largely established by age one—long before babies show any signs of knowing an abstraction from a pacifier.

Furthermore, studies are showing that spoken language has an astonishing impact on an infant's brain development. In fact, some researchers say the number of words an infant hears each day is the single most important predictor of later intelligence, school success and social competence. There is one catch—the words have to come from an attentive, engaged human being. As far as anyone has been able to determine, radio and television do not work.

"We now know that neural connections are formed very early in life and that the infant's brain is literally waiting for experiences to determine how connections are made," said Dr. Patricia Kuhl, a neuroscientist at the University of Washington in Seattle and a key speaker at the conference. "We didn't realize until very recently how early this process begins," she said in a telephone interview. "For example, infants have learned the sounds of their native language by the age of six months."

This relatively new view of infant brain development, supported by many scientists, has obvious political and social implications. It suggests that infants and babies develop most rapidly with caretakers who are not only loving, but also talkative and articulate, and that a more verbal family will increase an infant's chances for success. It challenges some deeply held beliefs—that infants will thrive intellectually if they are simply given lots of love and that purposeful efforts to influence babies' cognitive development are harmful.

If the period from birth to three years is crucial, parents may assume a more crucial role in a child's intellectual development than teachers, an idea sure to provoke new debates about parental responsibility, said Dr. Irving Lazar, a professor of special education and resident scholar at the Center for Research in Human Development at Vanderbilt University in Nashville. And it offers yet another reason to provide stimulating, high-quality day care for infants whose primary caretakers work, which is unavoidably expensive.

The idea that early experience shapes human potential is not new, said Dr. Harry Chugani, a pediatric neurologist at Wayne State University in Detroit and one of the scientists whose research has shed light on critical periods in child brain development. What is new is the extent of the research in the field known as cognitive neuroscience and the resulting synthesis of findings on the influence of both nature and nurture. Before birth, it appears that genes predominantly direct how the brain establishes basic wiring patterns. Neurons grow and travel into distinct neighborhoods, awaiting further instructions.

After birth, it seems that environmental factors predominate. One study found that mice exposed to an enriched environment have more brain cells than mice raised in less intellectually stimulating conditions. In humans, the inflowing stream of sights, sounds, noises, smells, touches—and most important, language and eye contact—literally makes the brain take shape. It is a radical and shocking concept.

Experience in the first year of life lays the basis for networks of neurons that enable us to be smart, creative and adaptable in all the years that follow, said Dr. Esther Thelen, a neurobiologist at Indiana University in Bloomington.

The brain is a self-organizing system, Dr. Thelen said, whose many parts cooperate to produce coherent behavior. There is no master program pulling it together but rather the parts self-organize. "What we know about these systems is that they are very sensitive to initial conditions," Dr. Thelen said. "Where you are now depends on where you've been."

The implication for infant development is clear. Given the explosive growth and self-organizing capacity of the brain in the first year of life, the experiences an infant has during this period are the conditions that set the stage for everything that follows.

In later life, what make us smart and creative and adaptable are networks of neurons that support our ability to use abstractions from one memory to help form new ideas and solve problems, said Dr. Charles Stevens, a neurobiologist at the Salk Institute in San Diego. Smarter people may have a greater number of neural networks that are more intricately woven together, a process that starts in the first year.

The complexity of the synaptic web laid down early may very well be the physical basis of what we call general intelligence, said Dr. Lazar at Vanderbilt. The more complex that set of interconnections, the brighter the child is likely to be since there are more ways to sort, file and access experiences.

Of course, brain development "happens" in stimulating and dull environments. Virtually all babies learn to sit up, crawl, walk, talk, eat independently and make transactions with others, said Dr. Steve Petersen, a neurologist at Washington University School of Medicine in St. Louis. Such skills are not at risk except in rare circumstances of sensory and social deprivation, like being locked in a closet for the first few years of life. Subject to tremendous variability within the normal range of environments are the abilities to perceive, conceptualize, understand, reason, associate and judge. The ability to function in a technologically complex society like ours does not simply "happen."

One implication of the new knowledge about infant brain development is that intervention programs like Head Start may be too little, too late, Dr. Lazar said. If educators hope to make a big difference, he said, they will need to develop programs for children from birth to three.

Dr. Bettye Caldwell, a professor of pediatrics and an expert in child development at the University of Arkansas in Little Rock, who supports the importance of early stimulation, said that in early childhood education there is a strong bias against planned intellectual stimulation. Teachers of very young children are taught to follow "developmentally appropriate practices," she said, which means that the child chooses what he or she wants to do. The teacher is a responder and not a stimulator.

Asked about the bias Dr. Caldwell described, Matthew E. Melmed, executive director of Zero to Three, a research and training organization

TIMETABLE: The Growing Brain: What Might Help Your Infant

Dr. William Staso, an expert in neurological development. suggests that different kinds of stimulation should be emphasized at different ages. At all stages, parental interaction and a conversational dialogue with the child are important. Here are some examples:

FIRST MONTH: A low level of stimulation reduces stress and increases the infant's wakefulness and alertness. The brain essentially shuts down the system when there is overstimulation from competing sources. When talking to an infant, for example, filter out distracting noises, like a radio.

MONTHS 1 TO 3: Light/dark contours, like high-contrast pictures or objects, foster development in neural networks that encode vision. The brain also starts to discriminate among acoustic patterns of language, like intonation, lilt and pitch. Speaking to the infant, especially in an animated voice, aids this process.

MONTHS 3 TO 5: The infant relies primarily on vision to acquire information about the world. Make available increasingly complex designs that correspond to real objects in the baby's environment; motion also attracts attention. A large-scale picture of a fork, moved across the field of vision, would offer more stimulation than just an actual fork.

MONTHS 6 TO 7: The infant becomes alert to relationships like cause and effect, the location of objects and the functions of objects. Demonstrate and talk about situations like how the turning of a doorknob leads to the opening of a door.

MONTHS 7 TO 8: The brain is oriented to make associations between sounds and some meaningful activity or object. For example, parents can deliberately emphasize in conversation that the sound of water running in the bathroom signals an impending bath, or that a doorbell means a visitor.

MONTHS 9 TO 12: Learning adds up to a new level of awareness of the environment and increased interest in exploration; sensory and motor skills coordinate in a more mature fashion. This is the time to let the child turn on a faucet or a light switch, under supervision.

MONTHS 13 TO 18: The brain establishes accelerated and more complex associations, especially if the toddler experiments directly with objects. A rich environment will help the toddler make such associations, understand sequences, differentiate between objects and reason about them.

for early childhood development in Washington, D.C., said that knowing how much stimulation is too much or too little, especially for infants, is "a really tricky question. It's a dilemma parents and educators face every day," he said.

In a poll released today, Zero to Three found that 87 percent of parents think that the more stimulation a baby receives the better off the baby is, Mr. Melmed said. "Many parents have the concept that a baby is something you fill up with information and that's not good," he said.

"We are concerned that many parents are going to take this new information about brain research and rush to do more things with their babies, more activities, forgetting that it's not the activities that are important. The most important thing is connecting with the baby and creating an emotional bond," Mr. Melmed said.

There is some danger of overstimulating an infant, said Dr. William Staso, a school psychologist from Orcutt, California, who has written a book called *What Stimulation Your Baby Needs to Become Smart*. Some people think that any interaction with very young children that involves their intelligence must also involve pushing them to excel, he said. But the "curriculum" that most benefits young babies is simply common sense, Dr. Staso said. It does not involve teaching several languages or numerical concepts but rather carrying out an ongoing dialogue with adult speech. Vocabulary words are a magnet for a child's thinking and reasoning skills.

This constant patter may be the single most important factor in early brain development, said Dr. Betty Hart, a professor emeritus of human development at the University of Kansas in Lawrence. With her colleague, Dr. Todd Ridley of the University of Alaska, Dr. Hart coauthored a book called *Meaningful Differences in the Everyday Experience of Young American Children*.

The researchers studied 42 children born to professional, working class or welfare parents. During the first two and half years of the children's lives, the scientists spent an hour a month recording every spoken word and every parent-child interaction in every home. For all the families, the data include 1,300 hours of everyday interactions, Dr. Hart said, involving millions of ordinary utterances.

At age three, the children were given standard tests. The children of professional parents scored highest. Spoken language was the key variable, Dr. Hart said.

A child with professional parents heard, on average, 2,100 words an hour. Children of working-class parents heard 1,200 words and those with parents on welfare heard only 600 words an hour. Professional parents talked three times as much to their infants, Dr. Hart said. Moreover, children with professional parents got positive feedback 30 times an hour—twice as often as working-class parents and five times as often as welfare parents.

The tone of voice made a difference, Dr. Hart said. Affirmative feedback is very important. A child who hears, "What did we do yesterday? What did we see?" will listen more to a parent than will a child who always hears "Stop that," or "Come here!"

By age two, all parents started talking more to their children, Dr. Hart said. But by age two, the differences among children were so great that those left behind could never catch up. The differences in academic achievement remained in each group through primary school.

Every child learned to use language and could say complex sentences, but the deprived children did not deal with words in a conceptual manner, she said.

A recent study of day care found the same thing. Children who were talked to at very young ages were better at problem solving later on.

For an infant, Dr. Hart said, all words are novel and worth learning. The key to brain development seems to be the rate of early learning—not so much what is wired but how much of the brain gets interconnected in those first months and years.

—Sandra Blakeslee, April 1997

Deaf Babies Use Their Hands to Babble, Researcher Finds

DEAF BABIES OF DEAF PARENTS babble with their hands in the same rhythmic, repetitive fashion as hearing infants who babble with their voices, a study has found.

The deaf babies, who presumably watch their parents use sign language at home, start their manual babbles before they are 10 months old, the same age hearing children begin stringing together sounds into wordlike units.

And just as hearing babies experiment with a few key noises like "dadadada" or "babababa," so deaf infants use several motions over and over, including one gesture that looks like "okay" and another that resembles a hand symbol of the numeral 1.

The gestures of the deaf children do not have real meaning, any more than babble noises have meaning, but they are far more systematic and deliberate than are the random finger flutters and fist clenches of hearing babies. The motions seem to be the deaf babies' fledgling attempts to master language, said Dr. Laura Ann Petitto, a psychologist at McGill University in Montreal. She is the principal author of the report, which appeared in the journal *Science*.

The new research strongly suggests that the brain has an innate capacity to learn language in a particular, stepwise fashion, by stringing together units into what eventually become meaningful words, Dr. Petitto said. The brain will progress from one stage to another regardless of whether language is conveyed through speaking, hand signing or presumably any other method of communication, she added.

The results contradict a widespread assumption among linguists that the maturation of the vocal cords affects language development among infants.

"For centuries, people thought that speech is language and language is speech," she said. "There's been a whole complicated notion that the structure of the motor apparatus and the unfolding of the mouth muscles actually influenced the structure and development of language."

But in showing that deaf babies babble with their hands in a manner that has all the basic elements of vocal babbling, she said, "We've decoupled language from speech. We've torn them apart."

Other researchers in the field of early language development praised the new study as significant and extremely well designed.

"I think this is important work," said Dr. Richard P. Meier, an assistant professor of linguistics and psychology at the University of Texas. "It's been suggested that all children pass through a regular sequence of milestones in speech acquisition, from simple cooing early on, to structured babbling at eight months, to the first word at about twelve months. This work gives us a new dimension of how language matures."

Dr. Marilyn M. Vihman, who is doing research on language acquisition at Rutgers University in New Brunswick, New Jersey, said the latest results offered further proof that babbling is a crucial step in language development. "You'll see the emergence of babbling at the same age in any infant who has been exposed to language regardless of other things, including intelligence," she said. "Babies babble regardless of whether the language spoken at home is English, Japanese, French or, it seems, sign language."

Dr. Petitto and her graduate student, Paula F. Marentette, videotaped five infants at ages 10, 12 and 14 months. Two of the infants were deaf children of deaf parents who use American Sign Language to communicate, while the other three were hearing offspring of hearing adults. The researchers analyzed every hand gesture of the infants and compared the two groups.

They found that the hearing children made many hand gestures, but that the gestures never became organized or repetitive. By comparison, the deaf babies soon began showing evidence of using about 13 different hand motions over and over again. Nearly all of them were actual elements of American Sign Language: gestures that do not in themselves mean anything, but have the potential to indicate something when pieced together with other gestures.

Sign languages are structured much like any spoken language, Dr. Petitto said. Distinct gestures and hand shapes are the equivalent of syllables, and thus must be presented in a series to assume any sense.

She believes that the deaf parents noticed the nascent efforts of their children to communicate through signs and began reinforcing the gestures, just as hearing parents talk back to and reinforce their babbling infants by turning the babble into words, for example, by saying "dadadada . . . Daddy."

But there was idiosyncratic taste at work as well. Just as one hearing baby may prefer to say "babababa" while another fastens upon "gagaga," so one of the deaf infants tended to make her gestures in front of her torso, while the other deaf baby performed his hand signs around his head and face.

The deaf babies could make noise, but they did not babble vocally like the hearing children.

Dr. Petitto said that the new work also supported the theory that the basic rhythm of all languages was the same, building upon a pattern that alternates consonants and vowels. Although the analogy is only approximate in sign language, the deaf babies did alternate between mobile gestures, which are thought to be somewhat vowel-like in their rhythms, and static hand shapes, the rough equivalent of the consonant.

—NATALIE ANGIER, March 1991

Removing Half of Brain Improves Young Epileptics' Lives

BRAIN-DAMAGED CHILDREN ARE ACTUALLY able to recover some intellectual ground if the entire damaged half of the brain is surgically removed, researchers are finding.

The surgical procedure, hemispherectomy, was first developed in the 1920s but fell out of favor for many years because of a high complication rate. Now newer surgical techniques have made the operation safer. Its success in children with damage confined to half the brain astonishes even seasoned scientists and suggests that until now, they may have greatly underestimated the brain's flexibility, particularly in older children.

"We are awed by the apparent retention of memory and by the retention of the child's personality and sense of humor," Dr. Eileen P. G. Vining of Johns Hopkins University wrote in the journal *Pediatrics*.

Dr. Vining reported on the progress of 54 children who underwent hemispherectomy for recurrent, severe epileptic seizures. Most of the patients stopped having incapacitating seizures after the operation and could stop taking high doses of antiseizure medication. Many are now in school.

The most extraordinary success of hemispherectomy was probably seen in a British boy named Alex who was born in 1980 with the left, speech-controlling, side of his brain smothered in a tangle of abnormal blood vessels, leaving him mute, half-blind, half-paralyzed and epileptic until he was eight.

Then Alex's doctors, unable to control his epilepsy with medication, removed the entire diseased left-brain hemisphere, warning his parents that although the seizures would be helped by the operation, his other problems would probably not change too much. Alex was well past the age when a mute child had ever been reported to learn to talk.

Ten months later, Alex startled everyone by suddenly speaking, first in words, then, a few months later, in complete sentences. At the age of 11 he was still pronouncing some words incorrectly "like a well-spoken foreigner," but now, at 16, he is extremely fluent, said Dr. Faraneh Vargha-Khadem, a neuropsychologist at University College in London who has followed Alex for six years and reported his case in the journal *Brain*.

Alex's case has given scientists "a new respect for the plasticity of the still-developing brain," said Dr. Mortimer Mishkin, a neuropsychologist at the National Institute of Mental Health in Maryland who has also studied Alex's progress.

Dr. John Freeman, the director of the Johns Hopkins Pediatric Epilepsy Center, said he was dumbfounded at the ability of children to regain speech after losing the half of the brain that is supposedly central to language processing.

"It's fascinating," Dr. Freeman said. "The classic lore is that you can't change language after the age of two or three."

But Dr. Freeman's group has now removed diseased left hemispheres in more than 20 patients, including three 13-year-olds whose ability to speak transferred to the right side of the brain in much the way that Alex's did.

"Speech transfers over at least to the age of thirteen, and possibly even older," Dr. Freeman said. "It's as if in the right hemisphere, all those words are there, but they're buried under tall grass and there aren't any pathways to get to them. And the older you are, the higher the grass is, and the harder the words are to find until the paths are formed. And memory is on both sides, too. When you take out one half of the brain, either half, these children have not forgotten anything."

Before hemispherectomy, Dr. Freeman said, some of the children followed at Johns Hopkins were having so many seizures that one episode blended into another in a state of continuous rhythmic jerking, despite high doses of medications that often left the children groggy.

In the eight-hour operation, as it is performed at Johns Hopkins, the brain tissue is removed a section, or lobe, at a time, said Dr. Benjamin Carson, a pediatric neurosurgeon there. The exposed surfaces of the remaining brain are insulated with a collagen-based material, and the empty space left in the skull eventually fills with cerebrospinal fluid.

After the operation, all the children had a very weak hand and arm on the side opposite the operation, blocked vision on that side and a weak leg that required a brace to walk or run. In many, however, strength and vision had already been compromised by the underlying disease, and the child actually became stronger after the operation rather than weaker.

Scientists still cannot explain why removing abnormal brain tissue may free the remaining tissue to function more normally. Getting rid of a child's seizures and taking a child off antiseizure medications, which can be powerful sedatives, may play a major role, Dr. Freeman said.

Dr. Vargha-Khadem, who has followed more than 100 children after hemispherectomy, offered this assessment of the procedure: "There's always a cost to brain injury. But if that brain injury occurs early, and if it is restricted to one hemisphere, and if the period of seizure disorder is not allowed to continue for too long, then the child really has got a fairly good chance to recoup a lot of the consequences of the brain damage and of the removal of the hemisphere."

—ABIGAIL ZUGER, August 1997

Test-Tube Moms

"GOD BLESS YOU, LITTLE BOY. Bless you." The mother's voice coos at her four-month-old, who is clearly pleased. "Oh, everybody's getting smiles now," the mother observes. "You're a cutie. . . ."

A soothing voice, a blessing, a cherubic smile. It has all the nuzzling intimacy of a quiet conversation between mother and child, except for the three-ton machine in the room and the fact that Mother's voice is canned, playing on a tape deck in the control room of the functional magnetic resonance imaging (FMRI) suite at the Memorial Sloan-Kettering Cancer Center. The maternal monologue, recorded several days earlier, is being piped into earphones wrapped around the head of the mother's very young son, who has been lightly sedated for medical tests.

In his short life, the child has already endured abdominal and brain surgeries for a rare congenital cancer. Now he is undergoing a postsurgical brain scan, and it is this diagnostic session that has provided the neuroscientist, Joy Hirsch, and her colleagues at Sloan-Kettering a rare scientific privilege: the opportunity to peer at a young human brain making first sense of the world. They are taking snapshots of a cognitive galaxy in the throes of formation.

In a series of meticulously plotted imaging experiments, each lasting exactly 198 seconds, Hirsch and her assistant stroke the boy's right hand with cotton gauze, flash lights into his closed eyes and play the tape of his mother's voice—all the while taking scans of his brain in action. The FMRI machine then sends a cataract of raw data down to Hirsch's lab for analysis.

"This is the youngest child we've ever done," confides Hirsch, a cheerfully voluble scientist whose Sloan-Kettering lab has conducted pioneering FMRI studies on developing brains.

While the study is still incomplete, it is possible to venture two predictions about the results. Hirsch's group will report provocative new findings on what babies know and when they know it, and everyone with a stake in the outcome—parents, educators and policy makers—will rush to incorporate those results in what has become a neurotic national pastime of late: raising a scientifically correct child.

It is precisely experiments like these, made possible in recent years by advances in new imaging technologies, that have inevitably left mothers, fathers and other caregivers reeling with the feeling that they must intervene early and often in the intellectual development of their children. Or else.

Matthew E. Melmed, executive director of Zero to Three, a not-for-profit child-development organization, has noticed "almost a sense of frenzy" among parents about the new neuroscientific research in the past few years. "While there's no question that the first few years of life are critically important, the message some parents are getting is, it's all over by the time their child is three years old, and it's not," he said.

The temptation to respond to every new finding about breast-feeding or language acquisition or mathematical skills is difficult for a concerned parent to resist. About the time our daughter was born two years ago, studies came out suggesting that the optimum period for imprinting a foreign language on a child comes to an end around the first birthday. This is known as a "critical period"—a rigid and unforgiving window of childhood development that slams shut at an early and biologically determined age.

My wife urged me to speak Italian to Micaela, and so I would occasionally jabber to her—*at* her, really—in my best *romanaccio*. In this proverbial, doughlike phase of life, she didn't have much choice but to look up with her big brown eyes and participate in the experiment. Now that she's verbal, mobile and independent-minded, her *pazienza*—as they say in the old country—has grown conspicuously short when Dad starts blabbering *Ciao*. In retrospect, I just feel silly about it.

At some level, every parent wants to be told what to do by "experts," but the flood of neuroscience findings has taken parental insecurity to a new level. In July 1997, the Sloan-Kettering group published a report in the journal *Nature* suggesting that people who learn a second language at a very early age use the same part of the brain in which they park knowledge of their native language, whereas people who acquire a second language later in life use a different part of the brain.

"We were barraged by letters from parents or school districts or individuals, demanding to know if it were a good thing or not to learn a second language at an early age," Hirsch recalled in her office.

She pulled out an enormous file of correspondence, which spoke volumes about the anxiety these studies elicit. Both sides of the current California debate on bilingual education contacted Hirsch as a potential ally. Someone planning an elementary school second-language curriculum wrote, "Do you have any idea at what age a child switches from the one pattern of brain usage for language to the other? If you've no data, do you have a guess?" And a writer from a parenting magazine asked, "When should parents begin teaching their child another language, and when would the cutoff be when the second language would start appearing in another part of the brain rather than in the same part as the first language?"

"Our results said nothing about the should question," Hirsch said. "We don't know. All we can say is that if you do learn a second language when you're young, our results show that it affects a specific neurological area."

And that is precisely the point: the results of experiments, undeniably exciting but nonetheless limited and preliminary, are being used by nonexperts to inform everything from the style of a mother's nurturing to the organization of day care centers to public policy on child rearing, with precious little scientific data to support ambitious, even aggressive leaps into future education policy. In some cases, neuroscience is being used to usurp the common sense of mothering.

A group of researchers at the Santa Fe Institute, for example, are trying to organize a community-based day care experiment that will incorporate recent neuroscience findings to improve the cognitive development of children up to three years of age. It is the group's belief that programs like Head Start begin a little too late. "We're talking about an experiment where we'd have fifty kids, and a control group of fifty kids," said George A. Cowan, the founding president of the institute. "Everyone has an opinion about babies, and there's the feeling that you don't need to do anything for zero to two-year-olds. You know, mothers are the ultimate arbiters, and babies just need tender, loving care. And that's true, but they need more than that."

But do they? A growing number of experts are beginning to make the point that it is far too early to take these findings as gospel.

"Neuroscience," said John T. Bruer, "doesn't know nearly as much as the public has been led to believe." Bruer is a philosopher of science by training and president of the St. Louis–based James S. McDonnell Foundation, which finances about $10 million in neuroscience-related research. So he was especially qualified to deliver a broadside against the popularization of neuroscience in a speech at Carnegie Mellon University. His remarks demolish a lot of misconceptions about what science can—and can't—tell us about a child's intellectual development.

According to Bruer, the problem isn't neuroscientists like Hirsch, whose experiments are illuminating early cognitive development. Rather it's in the way scientific information travels from the lab to journalists, educators, policy makers and ultimately to parents like me—a process in which details fall out and effects are magnified, resulting in an overreaction to preliminary findings. "It's not like there's a huge database," Bruer said. "What we're talking about is studies of twenty-five monkey brains here, about fifty human brains there, twenty-nine epileptic children there, and then building it into a kind of social-development theory of everything."

Dr. Floyd Bloom, a neuroscientist at the Scripps Research Institute and the editor in chief of the journal *Science,* expresses similar sentiments. "I don't think neuroscience has brought childhood development to the level where we can be the Dr. Spocks of the twenty-first century," he said.

The "neuroscience and education argument," as Bruer calls it, stands on three legs: synaptic development, enriched environments and critical periods. Shortly after birth, the brain undergoes an explosive increase in the number of synapses—the connections between brain cells. Early childhood experiences, it is argued, "sculpt" this mass of synapses. Synapses that are not wired into the brain's circuitry by early and repeated experience wither and disappear. Children deprived of stimulating experiences during this critical period may forever lose the opportunity to develop certain skills, like learning a language or developing logic. If "you miss the window," as *Newsweek* put it, "you're playing with a handicap."

That leaves a lot of parents with a scenario that goes something like this: the child listening to Bach will grow up to become a conductor or a mathematician, but the child playing in a sandbox is destined to spend life as a hod carrier. But of course it's much more complicated than that.

Why? For starters, most of the findings about the growth of synapses are based on studies of animals, not humans, and the research is difficult to interpret for a number of technical reasons, Bruer said.

Also, research shows that the creation and density of synapses may vary from one region of the brain to another, and that the number of synapses in these regions peaks at different times; moreover, some "critical periods" last well into adolescence. "Thus," Bruer said, "what neuroscientists know about synaptogenesis does not support a claim that zero to three is a critical period for humans."

The second supporting leg of the early-development argument derives from research on "enriched" environments. For nearly three decades, William T. Greenough and his colleagues at the University of Illinois at Urbana-Champaign have conducted a series of experiments on rats showing that those raised in "complex" environments—cages that contain many objects and toys—grow more synapses than those in cages without such furnishings. Those results have also been used to argue for the creation of enriched environments to enhance the cognitive development of young children.

Among the people most irritated about this leap is Greenough himself. "Some people are trying to make a growth industry by overinterpreting the data," he complained.

Finally, Bruer argued that the entire notion of critical periods, the third leg of the early-development argument, has been fatally simplified as it has been repackaged for popular consumption. Yes, research from the Greenough lab suggests that a developing neural system needs environmental cues and features to activate the brain, but nothing in the research suggests that reciting T. S. Eliot or declaiming in French is required, or even desirable.

"The expected experiences must be present during certain developmental periods," Bruer conceded, "but the expected experiences are of a very general kind—seeing visual patterns, the ability to move and manipulate objects, noises, the presence of speech sounds. These kinds of stimuli are available in any child's environment, unless that child is abused to the point of being raised in a sensory-deprivation chamber."

Moreover, the story clearly isn't over by age three; Greenough has shown that even elderly rats grow new synapses, as long as they live in complex, challenging environments. And a fairly recent finding is that, contrary to long-held wisdom, the brain grows new cells throughout life.

Meanwhile, Georgia's Democratic Governor, Zell Miller, mandated that every child born in the state be sent home from the hospital with a free classical music recording in the belief that this would improve mathematical aptitude. "It's tragic," Bruer said, "that we're basing childhood policy on a *Time* cover story."

What's the harm in playing "Eine Kleine Nachtmusik" to an infant? The harm has nothing to do with the philosophy of nurturing, and everything to do with money. "If zero-to-three programs are taking money from other programs," Bruer said, "that's something to worry about."

Michael Meaney, a neuroscientist specializing in early-childhood development at McGill University, made a similar point: "Head Start has come out with a list of predictors for what it calls life success. And they have nothing to do with your child knowing algebra at age three. They're social and emotional skills, like how they confront challenges and how they work with others."

A couple days after Joy Hirsch's team scanned the little boy's brain, I returned to her lab to see the results. Hirsch had been up until two in the morning preparing first-draft images of the data, and the preliminary pictures were spread out on a table before her.

The images of the four-month-old's brain, in addition to the pictures taken of other infants, showed subtle but consistent distinctions suggesting that very young preverbal children may indeed be capable of comprehension, and that the staging ground of early comprehension occurs in a specific place in the brain at a specific point in time. "It leads one to suspect," she said, "that we're looking at the headwaters of cognition."

"I suspect people will look at these results and say, 'Oh, how can I use this information to make my child better adapted to the challenges of adulthood?'" she said. "Those are questions that people have always been asking, those are questions they will continue to ask and of course we won't be able to provide any further answers. The time course of science is geological, while the time course of parenting is immediate. You can't wait ten years to get an answer." When your child is two years old, that's a long time to wait.

—STEPHEN S. HALL, April 1998

When an Adult Adds a Language, It's One Brain, Two Systems

AS THOUSANDS OF TEENAGERS who have struggled to engrave high school French on recalcitrant neurons might have guessed, it has been found that second languages are stored differently in the human brain depending on when they are learned.

Babies who learn two languages simultaneously, and apparently effortlessly, have a single brain region for generating complex speech, researchers say. But people who learn a second language in adolescence or adulthood possess two such brain regions, one for each language.

The findings, described in the journal *Nature,* shed new light on some notoriously difficult questions about brain development: how does the brain organize language in infancy and how are multiple languages represented in the brain? Why do some brain regions appear immutable after childhood, while others appear flexible and malleable in adult life? Why are languages harder to learn later in life?

There have always been strong hints that the brain can use separate brain regions for first and second languages, said Dr. Michael Posner, a psychologist at the University of Oregon in Eugene. Bilingual epilepsy patients may, during seizures, lose the ability to speak one language and not another. A stroke victim can permanently lose the ability to speak French but retain English or another language. "But it's not been known how these separate language areas form in the brain," Dr. Posner said. "Are the languages fused? Do they prime one another? Is one translated by another?"

The new study shows for the first time that two languages can be mapped in common neural tissue, Dr. Posner said, adding, "It is very helpful for understanding bilingualism."

The research was carried out by Dr. Joy Hirsch, head of Memorial Sloan-Kettering Cancer Center's functional MRI Laboratory and her graduate student, Karl Kim. Functional magnetic resonance imaging, or MRI, is a relatively new, noninvasive brain imaging technique that can pinpoint exactly which parts of the brain are active during cognitive tasks such as talking, seeing, waving an arm or daydreaming. Brain surgeons at the hospital are now using the technique to identify critical brain regions so that they will not do more harm than good when removing a tumor or other abnormality.

Of these critical regions, language is perhaps at the top of the list, said Dr. Philip H. Gutin, the hospital's chief of neurosurgery. Some functions such as seeing and hearing are located in both brain hemispheres, he said. When a tumor forms, surgeons can cut out tissue and not do great harm because the other side of the brain will take over. "But language is a high-rent district," Dr. Gutin said. Some high-level aspects of language tend to be found only on one side of the brain. By removing a spot of tissue smaller than an eraser, a surgeon could excise a crucial region of language production and destroy a person's ability to speak or understand English.

Moreover, language areas are never found in exactly the same spot, Dr. Gutin said. These regions are formed in childhood as language is acquired and are in slightly different spots in different people. Given that a quarter of all brain tumors occur in regions of the brain where language skills might reside, accurate imaging is a must, he said.

To explore where languages lie in the brain, Dr. Hirsch recruited 12 healthy bilingual people from New York City. Ten different languages were represented in the group. Half had learned two languages in infancy. The other half began learning a second language around age 11 and had acquired fluency by 19 after living in the country where the language was spoken.

With their heads inside the MRI machine, subjects thought silently about what they had done the day before using complex sentences, first in one language, then in the other. The machine detected increases in blood flow, indicating where in the brain this thinking took place.

Aspects of language ability are distributed all over the brain, Dr. Hirsch said. But there are some high-level, executive regions that are usually localized in a certain neighborhood on the left side of the brain, but are sometimes found in the same neighborhood on the right side, or on both sides. One is Wernicke's area, a region devoted to understanding the meaning of words and the subject matter of spoken language, or semantics. Another is Broca's area, a region dedicated to the execution of speech as well as some deep grammatical aspects of language. The regions are identified by observing brain function.

None of the 12 bilinguals had two separate Wernicke's areas, Dr. Hirsch said. In an English and Spanish speaker, for instance, Spanish semantics blended with English semantics in the same area. But there were dramatic differences in Broca's areas, Dr. Hirsch said.

In people who had learned both languages in infancy, there was only one uniform Broca's region for both languages, a dot of tissue containing about 30,000 neurons. Among those who had learned a second language in adolescence, however, Broca's area seemed to be divided into two distinct areas. Only one area was activated for each language. These two areas lay close to each other but were always separate, Dr. Hirsch said, and the second language area was always about the same size as the first language area.

This implies that the brain uses different strategies for learning languages, depending on age, Dr. Hirsch said. A baby learns to talk using all faculties—hearing, vision, touch and movement—which may feed into hardwired circuits like Broca's area. Once cells in this region become tuned to one or more languages, they become fixed. If two languages are acquired at this time, they become intermingled.

But people who learn a second language in high school have to acquire new skills for generating the complex speech sounds of the new tongue, which may explain why a second language is harder to learn. Broca's area is already dedicated to the native tongue and so an ancillary Broca's region is created. But Wernicke's area, which handles the simpler semantic aspects of language, can overlap.

—SANDRA BLAKESLEE, July 1997

Old Brains Can Learn New Language Tricks

PITY THE JAPANESE TOURIST ASKING for directions in New York City: "Which way to Times Square?"

The answer might be: "Turn left after the next light."

The next what? Does that mean traffic signal or the next street turning off to the right? To the typical native speaker of Japanese, right and light are hopelessly confused because the English sounds "l" and "r" are indistinguishable.

But they will not be confused for long. In a fascinating set of experiments, researchers at the Center for the Neural Basis of Cognition in Pittsburgh have found a way to teach native speakers of Japanese to hear the difference between "l" and "r" reliably after just one hour of training.

The findings are "extremely compelling," said Dr. Edward Jones, a neuroscientist at the University of California at Davis and president of the Cognitive Neuroscience Society, who is familiar with the research. They shed light on how the adult brain changes, a phenomenon called "plasticity," and on mechanisms that make it resistant to change.

Dr. Helen Neville, a leading expert on brain plasticity at the University of Oregon in Eugene, called the experiments "cool." They show that the adult brain is capable of substantial change, even late in life, she said.

The research—conducted by co-director Dr. Jay McClelland and his colleagues at the Pittsburgh center, a joint program of Carnegie Mellon University and the University of Pittsburgh—has long been interested in how the brain learns and, sometimes more important, how it fails to learn. He presented his findings at the annual meeting of the Cognitive Neuroscience Society, which was held in Washington.

The brain presents a puzzle, Dr. McClelland said, in that many kinds of learning continue or even improve throughout adulthood, but others, like the speech sounds of a language, appear to slow almost to a stop. Few people can learn a second language without an accent after the age of 10. Scientists call this 10-year window a critical period for acquiring the sounds of a language. But what is the neural basis of this critical period?

Clues are found in the way brain cells connect and influence one another. In test tube experiments, scientists take two nerve cells that are connected by a kind of cable called an "axon." When the first cell is induced to fire an electric pulse down its axon, the second cell also fires. Soon, physical changes develop in both cells so that the first one almost always makes the second one fire. This is how information is passed throughout the brain in a process known as "Hebbian learning."

Early on, brain cells are not well connected, Dr. McClelland said. But when experiences from the outside world begin to flow into the brain, cells begin to fire, and Hebbian patterns get stamped in. As neuroscientists are fond of saying, cells that fire together, wire together, forming circuits.

Later, a cell may come into direct contact with cells from another circuit, but if it is committed to what it has learned, it will not respond even to very strong stimulation from those other cells. In other words, it fails to learn.

This model of Hebbian learning can be applied to the sounds of human language, Dr. McClelland said. Newborn babies can discriminate all the sounds of every language in the world. It is as if there were a space inside their brains that is a blank slate, waiting for sounds to enter. When the sounds of the native language come pouring in, each sound induces some cells to wire up and become dedicated to its peculiar frequency.

Thus, there are clumps of cells that are tuned to the Finnish "o," the Spanish "d" or the English "th"—all of which are difficult for nonnative speakers to hear or pronounce. Within a short time, the baby's ability to distinguish all sounds fades away.

Babies in English-speaking families have cells dedicated to hearing both "l" and "r," whereas Japanese babies have only one phonetic category for a similar sound, Dr. McClelland said. To an American, the single Japanese "r" sound, as in the word *riokan,* meaning "guest room," sounds rather like a "d," he said. In any case, as the children in both cultures grow up, their sound categories become more sharply defined.

The brain has a relatively small amount of neural tissue dedicated to speech sounds, Dr. McClelland said, and so carves up that space with strong boundaries.

Sounds are stamped in early because the baby needs them to build the foundation for language comprehension. Thus, Hebbian processes come in early and lock in speech sound circuits. In other parts of the brain, Hebbian processes continue, but those circuits can be more flexible, he said. An adult learns to speak a second language by making new connections in many circuits but cannot supplant the locked-in native sound system with the new sounds of the second language.

Moreover, the unfamiliar sounds of a foreign language actually reinforce the sound system of one's native language, Dr. McClelland said. When a Japanese speaker comes to America, every time he hears an English "l" or "r," his single Japanese "r" phoneme is activated. Instead of becoming more flexible, his ability to hear "l" and "r" diminishes with increasing exposure to English.

The challenge, Dr. McClelland said, is to carve out two spaces in the Japanese speaker's brain when he has only one space for the "l" and "r" sounds. If the critical period is over forever, he said, this should be impossible. But if some plasticity or malleability remains, there should be a way to override the embedded circuits using Hebbian learning.

Thirty-four native Japanese speakers came to the Pittsburgh laboratory, where they were given headphones through which they heard pairs of words—road and load, light and right, and so on—under one of two conditions.

In one condition, subjects heard regular speech. They had to say or guess when they heard an l-word or r-word by tapping a response into a laptop computer.

In the second condition, subjects heard the same words exaggerated by a computer, so that each sound's peculiar frequency or formant was accentuated. As their ability to distinguish l- and r-words improved, the words were presented in regular speech. Finally, they heard the words in sloppy or degraded speech so that even native speakers would have to listen hard to hear the difference.

The subjects, all of whom had great trouble with "l" and "r" before training, used the computer for three 20-minutes sessions, involving hundreds of word pairs, Dr. McClelland said. No feedback was offered, to make conditions resemble the real world.

Those who heard natural speech barely improved and some actually got worse, Dr. McClelland said. But those who heard the exaggerated speech with gradual training toward more natural speech all improved greatly. After one hour, they could clearly distinguish "light" from "right."

At this point, the successful subjects do not generalize what they have learned to all "l" and "r" sounds, he said. But the experiment is just beginning. If they go on to train on numerous pairs of words, they may be able to retrain their entire sound system. It appears that they have begun to carve out new, independent circuits for the "l" and "r" sounds.

This approach may be effective for retraining other embedded circuits, like those that underlie racial prejudices or stereotypes, Dr. McClelland said. For example, some people react with fear when they see a stranger, based simply on his dress or skin color. That response may be stamped in. "We could think about ways of structuring situations to present a stimulus that would originally elicit the fear response and then teach the brain to have a different reaction," he said.

And it can definitely be used for second language training.

During the World War II, American soldiers fighting the Japanese infantry adopted the password "lollapalooza," figuring that no Japanese speaker could pronounce it. So much for that idea.

—SANDRA BLAKESLEE, April 1999

Scientists Track the Process of Reading Through the Brain

SCIENTISTS STUDYING THE PROBLEM OF dyslexia have found a neural pathway for reading. People who have great difficulty reading, they discovered, use that pathway only inefficiently as they struggle to make sense of written words.

The researchers, Dr. Sally E. Shaywitz and Dr. Bennett A. Shaywitz, the co-directors of the Yale Center for Learning and Attention at Yale University, drew on decades of research that broke the process of reading down into its fundamental tasks. Then, asking volunteers to perform a hierarchy of these tasks, they imaged the brain to see which areas were active. They discovered, Dr. Sally Shaywitz said, that poor readers, people with dyslexia, have a "glitch in the wiring" of the brain's pathway that is used for reading.

Although there is a popular notion that people with dyslexia reverse letters, that in fact, is not the essence of the disorder. Instead, researchers say, dyslexia is defined as great difficulty in reading that is not explained by a lack of intelligence.

The Shaywitzes, a married couple, and others said that the new work can be used to discover better ways of diagnosing and treating reading disorders. And it can be used to study children when they are learning to read, allowing investigators to discover whether the brain's pathways are modified as people learn to read and, if so, if that process can be facilitated in people with dyslexia.

The work was published in *The Proceedings of the National Academy of Sciences*.

Dr. Reid Lyon, chief of the child development and behavioral branch at the National Institute of Child Health and Human Development, noted that as many as one in five American children have great difficulty reading. And, he added, even if these children learn to read and are at least average readers, they never read for pleasure. "They don't like to read because it is still too hard for them," Dr. Lyon said.

And so, for decades, investigators have asked why some people find reading so hard.

"Over time, what we've tried to do is figure out a way to work with these kids," said Dr. Sheldon Horowitz, director of professional services at the National Center for Learning Disabilities in New York. But, he added, "previously we were not able to say with confidence that there was a neurological basis" for dyslexia. The notion persisted among parents, some educators and some reading experts that children with dyslexia were unintelligent, did not pay attention or did not try hard enough.

In the meantime, various neurobiologists and psychologists posited the theory that dyslexia involved disruptions in specific areas of the brain, Dr. Horowitz said. Some said dyslexia was a problem with the eyes—what people see when they try to read. Others, he added, thought it involved problems with hearing the sounds of language. But, Dr. Horowitz said, there was no common pathway and no way to make sense of the disparate hypotheses.

And so, Dr. Horowitz said, the Shaywitzes' work, which shows a specific pathway in the brain that good readers use but poor readers do not, is "incredibly exciting."

Dr. Lyon also praised the work. He said the study was "the first to demonstrate that there is a neurological system that undergirds all the components of reading and that a neurobiological signature underlies poor performance on reading."

The idea of breaking down reading into its component parts relied on decades of studies demonstrating that reading requires an ability to match letters with the sounds that represent them. And it also requires phonological awareness, or an ability to break words into their component sounds. Dr. Lyon explained: "If I say the word, 'bag,' how many sounds do you hear? Most people say three. But when you hear the word spoken, you hear one pulse of sound, not three. The ear doesn't hear those sounds in linear sequence—they are all mushed together and the brain has to pull them out." People who have trouble reading, Dr. Lyon said, "can't recover the sounds from speech."

Making use of the spectrum of tasks involved in going from letters to sounds to words, the Shaywitzes designed a hierarchical series of reading tasks and asked 29 adults with dyslexia and 32 who read easily to perform them. While the subjects attempted the tasks, the Shaywitzes watched their brains in action with functional magnetic resonance imaging, which shows differences between blood carrying oxygen and blood that is depleted of oxygen. An active area of the brain uses fresh supplies of oxygen-rich blood.

First the subjects had to recognize letters, the first step in reading. Then they were asked to decide whether letters like g and c, or h and p, rhymed. Then they had to decide whether nonsense words, like geat and lete, rhymed, a task that required that readers sound out the words.

As people moved from one task to the next, harder one, the Shaywitzes watched the pathways of their brains light up. In normal readers, that pathway involved fingertip-size regions on the surface of the brain and moved from the back of the brain to the front.

The path starts with the primary visual cortex. This area registers what the eyes see. Then an area called the "visual association area," or "angular gyrus," takes over. This is a region of the brain that translates the abstract scrolls of words and letters into language. The final area, behind the eyes and toward the front of the brain, is the superior temporal gyrus, or Wernicke's area. Here the brain takes the sounds of language and converts them into words.

People with dyslexia, the Shaywitzes found, barely used this reading pathway in the brain. Instead, another area of their brains lit up—the inferior frontal gyrus, or Borca's area. This region, toward the front of the brain, pairs words with units of sound.

The Shaywitzes noted that the angular gyrus was long suspected to be required for reading. A century ago, doctors noticed that when people had strokes or brain damage that destroyed this piece of the brain, they could no longer read. But it can be tricky to infer brain functions from such chance findings because the destroyed areas may be only indirectly affecting a function like reading.

The next step is to ask what happens when a child who has difficulty learning to read is taught to read with some facility. Do the brain's pathways change? Or does the child learn to use an alternate pathway? And what might that pathway be?

The Shaywitzes are now starting a study to answer those questions.

Dr. Judith Rumsey, a psychologist at the National Institute of Mental Health in Bethesda, Maryland, who studies the neurological basis of dyslexia, said that studies with children who are learning to read might also reveal prognoses. It might be, she said, that some children with dyslexia eventually learn to use the normal pathway for reading but others who, perhaps, are more severely affected, do not.

"Maybe you could see that on a scan to predict who is going to get better and who is not," Dr. Rumsey said. The scans might also be used to decide which teaching methods might be most useful for different groups of children with dyslexia. It might be possible to decide which children would benefit from continued instruction in reading and which children would be unable to learn to read, no matter how well they were taught. For these children, Dr. Rumsey said, "You might say, 'Why don't you rely on talking books and readers?'"

"I've seen a lot of adults who are horrible readers and they are smart people," Dr. Rumsey said. "It may well be that some people are just never going to get the knack for reading."

—GINA KOLATA, March 1998

4

LANGUAGE AND THE BRAIN

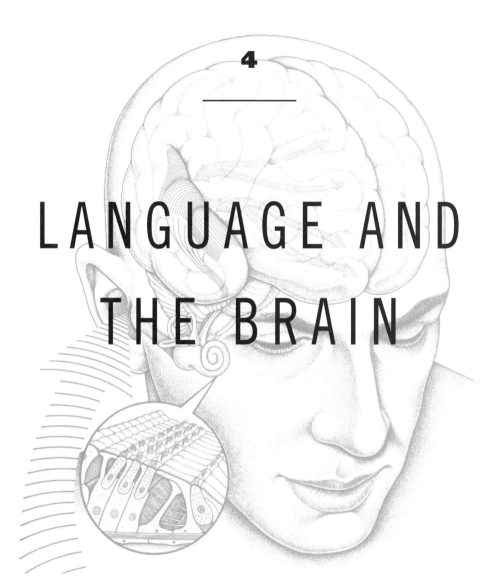

Textbooks long taught that language was confined to two specific areas on the surface of the cerebral cortex—Broca's area, where language was produced, and Wernicke's area, where it was comprehended. New research has showed this assertion to be a gross oversimplification, but there is not yet a coherent explanation to replace it.

What does seem clear is that the brain's basic language system allows for considerable variation in its execution. The brains of men and women, for example, handle language differently. A second language tends to occupy a more diffuse area of the brain than a person's first language, perhaps because it has to compete for neurons already committed to some other task.

The brain does not seem to possess a single dictionary that is the sole source of definitions. Rather, the meanings for different parts of speech, like proper nouns, common nouns and verbs seem each to have their own lexicons. There is even some evidence that the lexicons for speaking differ from those for comprehension.

Language is the most recent acquisition of the primate brain and perhaps the most sophisticated. So it will not be surprising if the mental systems that underlie language turn out to be highly complex. This ability we accomplish so easily depends on a series of skills, each one a minor miracle of biological engineering. The brain must decode the stream of electrical signals it receives from the ear into words. The words must be parsed, as the brain figures out the grammatical structure of the incoming sentence, and also married to the right meaning by being compared to the appropriate lexicon.

Neuroscientists are far from understanding how these separate tasks are accomplished, or the wizardry that wires them together in a system that operates so perfectly, apart from the occasional stutter or malapropism. An array of new tools, from genetic analysis to magnetic resonance imaging, is now helping to open new windows into the brain's abilities. Despite the distance that remains, considerable strides have been made in understanding how the brain is organized to comprehend and generate language.

Brain Yields New Clues on Its Organization for Language

CHRISTOPHER, 29 YEARS OLD, CANNOT draw simple figures, add 2 and 2 or tie his shoes. Yet he speaks 16 languages, half of them fluently.

Adam, 28, is a stroke patient. He can name man-made objects like saws, screwdrivers and shovels, but the stroke has left him unable to name most animate objects. The curious result is that ducks, foxes, camels and zebras are indistinguishable to him.

Carla, 22, grew up speaking both her native Italian and English. When she began training to become a simultaneous translator, her language ability was localized on the left side of her brain. But after the training, English shifted to her right brain while Italian remained on the left.

Study of special individuals like these, together with sophisticated new instruments like PET scanners, have afforded a startling new insight into how the brain is organized to handle language. Neurophysiologists are beginning to suspect that there is not a single center for language, but rather that the brain distributes language processing over some or many areas. The finding has prompted new ideas about how the brain thinks, including a provocative theory that proposes the brain has "language convergence zones" in which it brings together the separately located attributes of a word or concept.

"The time is ripe for an attack on the neural basis of consciousness" and language, says Dr. Francis Crick, the famous molecular biologist turned neuroscientist. Attempts to infer the nature of language through psychological experiments will never suffice, says Dr. Crick, who works at the Salk Institute in La Jolla, California. The problem can only be solved by explanations at the neural level.

Among the most interesting new findings about the brain and language are the following:

- Language is not located where people previously thought it was. Rather, each individual has a unique brain pattern underlying his or her language ability.
- Like the Cray computer, a person's first language is tightly organized in terms of nerve cell circuits. Second languages are more loosely organized in the brain, which is why it often takes longer to find words in them. But a stroke in one part of the brain can knock out a native language and leave later-learned languages intact—or vice versa.
- Different aspects of language, like proper nouns, common nouns and irregular or regular verbs, are processed in different areas of the brain. But these areas do not send their signals to a common destination for integration, as if language appeared on a Cinemascope screen in the brain. Rather, language and perhaps all cognition are governed by some as yet undiscovered mechanism that binds different brain areas together in time, not place.

The most direct insights into the brain's organization for language come from neurosurgeons who take the opportunity to record nervous activity in patients' brains during operations. By mapping the location of nerve cells that produce language, the neurosurgeons are revolutionizing established notions of how language is organized in the brain.

The traditional view holds that spoken and written language are processed in two structures, Broca's and Wernicke's areas, found on the left side of the brain. The right side of the brain was believed to handle spatial tasks and not to be involved in language.

But the brain's language areas are not so neatly compartmentalized, said Dr. George A. Ojemann, a neurosurgeon at the University of Washington in Seattle who is a leading brain mapper. While it is true that most people have essential language areas on the left side of their brains, he said, some people have them on the right side and others on both sides.

Even more surprising, Dr. Ojemann said, is that each person appears to have a unique pattern of organization for language ability—as unique as facial features or fingerprints. Broca's and Wernicke's areas are indeed important language-processing regions in most people, he said, but many additional language-areas are found elsewhere in the brain. Two left-brain regions called the temporal and parietal lobes are particularly rich in multiple-language areas, he said.

Each essential language area is composed of a sharply defined patch of nerve cells, each about the size of a grape, Dr. Ojemann noted. The cells in each patch appear to be connected to many others located in distant parts of the brain. Different patches govern language functions such as reading, identifying the meaning of words, recalling verbs and processing the words and grammars of foreign languages, he said.

Dr. Ojemann maps language areas in the brains of conscious individuals whose skulls have been opened prior to surgery. In one technique he maps the language-associated area by sticking electrodes into nerve cells on the exposed surface of the brain. In another, he inserts electrodes at random into cells deep within the brain so as to identify the distant nerve cells, or neurons, involved in naming objects.

After probing hundreds of brains, Dr. Ojemann has found that essential areas for naming things in one's native language are more compactly organized than those for later learned languages. Second languages tend to be diffusely organized, he said, as if neurons devoted to the new language were competing for space in existing essential areas. But when people are gifted bilinguals, Dr. Ojemann said, the brain develops separate, tightly organized essential areas for naming in each language. The same must be true for all language-essential areas, he said.

Brain mapping also shows sex differences in language ability, Dr. Ojemann said. Men tend to have larger essential areas in the parietal lobes, as do women with lower verbal I.Q.'s, he said. Men and women with high verbal I.Q.'s tend to have naming sites in an area called the middle temporal gyrus. Dr. Ojemann's research is described in the *Journal of Neuroscience*.

The process of learning a language undoubtedly shapes the formation of the essential areas, said Dr. Elizabeth Bates, a professor of psycholinguistics and an expert in child language acquisition at the University of California in San Diego. From birth to the age of two, she said, the child's brain undergoes an explosive growth of synaptic connections and is primed to learn the sounds and grammar of any language. After the age of two, she said, language synapses that do not receive inputs from early vocalizations begin to be eliminated or suppressed—a process that continues until about age 15.

Thus Japanese newborns can distinguish the sounds "ra" and "la" but by age two have begun to lose that ability, Dr. Bates said. Most young children can learn two or more languages without an accent, she said, but most people lose that ability by young adulthood.

Nevertheless, the adult brain can undergo dramatic changes associated with languages, according to Dr. Franco Fabro, a researcher at the University of Trieste in Trieste, Italy. Several years ago Dr. Fabro conducted a study of simultaneous translators, including a young woman named Carla. At the beginning of Carla's training, he said, English and Italian were strongly lateralized in her left brain, as judged by advanced electrical imaging techniques. But after training, the English appears to have shifted to the right side of her brain, he said, perhaps to avoid competing for essential language areas on the left, and Carla was able to shift instantly from language to language.

Dr. Ursula Bellugi of the Salk Institute and Dr. Edward Klima of the University of California at San Diego have studied the rich but silent languages of deaf people. Since these are based on hand movements in three-dimensional space and since the right brain is specialized for tasks involving spatial relationships, Dr. Bellugi wondered if sign language would prove to be localized in the right hemisphere. From deaf people who had suffered strokes, she found that it was lesions to the left brain that impaired the ability to sign. The left brain is strongly predisposed for language, Dr. Bellugi concluded.

Given these biological properties of language, is there a newly evolved organ that gives humans the power to generate language, as the MIT linguist Noam Chomsky and his followers say, or does language ability stem from the general properties of the brain and hence is widely distributed? An intense debate is under way.

One of the most fascinating pieces of evidence that bear on the debate is the case of Christopher, an idiot savant whose remarkable linguistic abilities support the idea of language as a separate organ. Christopher is socially inept, avoids eye contact and has a nonverbal I.Q. of 65, said Dr. Neil Smith, a linguist at University College London in England. Christopher cannot draw simple figures, carry on a very long conversation nor care for himself in the everyday world. Yet he has learned 16 languages and is a gifted translator.

Dr. Smith recently tested Christopher's grammatical knowledge of foreign languages and found him on a par with gifted polyglots. When Christopher met a speaker of Berber, a new language for him, he asked if the script could be written in Tifanagh, said Dr. Smith. It is a medieval script used by Berber women for writing love poetry.

But further tests reveal that Christopher's language abilities are independent of his cognitive abilities, Dr. Smith said. He never mulls over the meaning of passages and is not able to think about what he translates, he said.

Christopher's case suggests that language is processed in a separate brain organ that can remain intact in a damaged brain, said Dr. Vicky Fromkin, a Chomskyite linguist at UCLA. Conversely, when the language organ is damaged by stroke or other injury, she said, certain aspects of language are permanently lost—as when English speakers with Broca's aphasia lose the ability to use articles of speech: connectives and so-called function words as opposed to nouns and verbs.

Dissenting from the Chomskyite view are scientists who have found that computer networks with randomly assembled connections can learn to categorize words grammatically and produce correct English sentences. The computers even make "speaking" errors just as people do, said Dr. Bates. The finding suggests that a specialized computer or organ is not necessary for learning a language, she said, because a computer with no special structure can be taught.

Also, cross-cultural studies of the loss of language ability suggest that language and grammar are distributed widely across the brain, said Dr. Bates. The same Broca's area lesion that leads English speakers to leave out articles 70 percent of the time, she said, will cause Italian speakers to omit articles 25 percent of the time and German speakers to omit articles 5 percent of the time; in each language articles have different functions.

"This is because articles in English do not carry much meaning," Dr. Bates said, "whereas in Italian they carry information about number and gender and in German the article carries the case, telling you who did what to whom." A German speaker with the identical lesion will struggle to find the article because it is so important, she said, suggesting that knowledge of the language is not gone.

Actually, both sides are correct, said Dr. Antonio Damasio, a neurologist at the University of Iowa. With his wife, Dr. Hannah Damasio, he has developed a theory of cognition and language that is receiving rave reviews from linguists and neuroscientists, including Dr. Bates and Dr. Fromkin, who say they rarely agree on anything.

The brain has special areas for processing language very much along the lines of Dr. Ojemann's essential areas, Dr. Damasio said, but these areas do not constitute an independent language organ with little boxes where nouns, verbs and other language features are processed.

Rather, the essential areas can be thought of as "convergence zones" where the key to the combination of components of words and objects is stored. Thus knowledge of words and concepts is distributed widely throughout the brain but needs a third-party mediator—the convergence zone—to bring the knowledge together, during reactivation.

An example helps explain this difficult concept, Dr. Damasio said. "When I ask you to think about a Styrofoam cup," he said, "you do not go into a filing cabinet in your brain and come up with a ready-made picture of a cup. Instead, you compose an internal image of a cup drawn from its features. The cup is part of a cone, white, crushable, three inches high and can be manipulated.

"In reactivating the concept of this cup," Dr. Damasio said, "you draw on distant clusters of neurons that separately store knowledge of cones, the color white, crushable objects and manipulated objects. Those clusters are activated simultaneously by feedback firing from a convergence zone. You can attend to the revival of those components in your mind's eye and from an internal image of the whole object."

The same process is true of words, Dr. Damasio said. "When I ask you to tell me what the object is, you do not go into a filing cabinet where the word 'cup' is stored," he said. "Rather, you use a convergence zone for the word 'cup' by activating distant clusters of neurons that store the phonemes 'c' and 'u' and 'p.' You can perceive their momentary revival in your mind's ear or allow them to activate the motor system and vocalize the word 'cup.'"

To read, speak or make other lexical operations about a Styrofoam cup, the brain requires a third-party convergence zone that mediates between word and concept convergence zones, the scientist said. "Only then can we operate linguistically and evoke the word from the concept or vice versa," Dr. Damasio said.

Convergence zones are probably established in early childhood during language learning and as memories are formed, Dr. Damasio said. New ones are formed and old ones can be rearranged throughout life.

The convergence zone concept explains the odd language disabilities of his stroke patient Adam, Dr. Damasio noted. When shown a picture of a dog, Adam can say it is man's best friend, has four legs and barks—but he cannot summon the name for dog, Dr. Damasio said. Nor can he distinguish one animal from another by its name. But Adam can name different man-made tools with ease. The explanation: language convergence zones for natural objects are significantly damaged, but zones for man-made objects are largely intact.

Recent studies using a powerful brain imaging technique called PET, positron emission tomography, support the idea of convergence zones, said Dr. Steven Petersen of Washington University of in St. Louis. Some brain lesions, locatable with PET images, prevent people from reading while other language abilities remain intact, he said. Thus a person can speak the word cup, read the letters "c" and "u" and "p," trace the word "cup" and if the word is dictated, he or she can write it, Dr. Petersen said. But a little while later, they cannot read it, he said, even in their own handwriting. The area damaged is a zone for higher order processing of visual word-like forms, he said, possibly a convergence zone for reading.

One critical question about convergence zones remains unanswered. What makes populations of widely distributed neurons activate simultaneously? Dr. Damasio says he believes that the system mechanism is the feedback firing from a convergence zone, but the microscopic nature of the mechanism is a mystery. Many leading scientists, including Dr. Crick, think that the brain creates unified circuits by oscillating distant components at a shared frequency.

—Sandra Blakeslee, September 1991

The Mystery of Music: How It Works in the Brain

A 61-YEAR-OLD CANADIAN BUSINESSMAN, who asked that his name not be used, recently took his wife to a fancy restaurant to celebrate their wedding anniversary. As they sat down, he asked the musicians to play the couple's favorite song, "La Vie en Rose."

When the meal ended, the man turned to his wife and said, "I'm so sorry, darling. They didn't play our song."

She looked back sympathetically and said, "But they did play it. Three times."

Neither of them was really surprised. Since suffering a stroke 15 years ago, the man has been afflicted with a rare condition called amusia. He cannot recognize any music or songs, however familiar they once were, though his speech and other auditory faculties are mostly normal. He can hear music, tap out the rhythm and dance to it, and even respond to it emotionally. But the music sounds weird and distorted, and he cannot distinguish Beethoven from Chuck Berry.

Patients like this man are helping neuroscientists plumb the mysteries of music in the human brain, raising questions whose answers are as deep and intricate as a Bach cantata:

How is music perceived by the brain, and which cells and circuits come into play?

Has music blossomed in the march of human evolution, becoming a uniquely human trait, or are other animals, such as songbirds, equally musical?

What is the relationship between language and music?

Why does music tap our emotions?

What makes a tune stick in our heads? What makes some people more musically talented than others? Are musicians' brains wired differently from other people's?

And, finally, a question asked by many parents: how does music influence a developing child's brain?

While many answers are being found with the help of brain imaging machines and experimental techniques, the question of how and why music arose in human evolution remains speculative.

The ability to perceive and enjoy music is an inborn human trait, said Dr. Mark Tramo, a neurobiologist at Harvard Medical School. While many animals use intricate sounds to recognize one another, attract mates and signal danger, humans have developed the richest musical repertoires of any species.

A crowning achievement of human evolution is the ability to communicate complex ideas and emotional states, Dr. Tramo said. The human brain has evolved specialized circuits, called "feature detectors," for this purpose, which can be used to decode aspects of both speech and music. For example, the temporal lobes at both sides of the head contain cells that recognize and process pitch, which is the unitary pattern of frequencies that one hears when a musical instrument or the vocal cords are vibrated.

"Pitch is a part of grammar," Dr. Tramo said. "When you ask a question, the pitch rises. When you string individual speech sounds and sequences into sentences to express ideas, you begin to use pitch."

Pitch is also an integral part of music. When the brain listens to music, Dr. Tramo said, it uses many of the same pitch detectors that are used in decoding spoken language.

The first musical instrument was probably the human voice, Dr. Tramo said. As language flourished, so did music, with cultures inventing different kinds of resonators—flutes, reeds, simple strings. While language was used to transmit knowledge, he said, music was used to promote social cohesion through shared tribal rituals.

Some scientists think that language and music are two sides of the same intellectual coin, a view supported by the anatomical distribution of feature detectors in the human brain. Cellular circuits that recognize language and music are found on both sides of the brain, said Dr. Jamshed Bharucha, a psychologist at Dartmouth College in Hanover, New Hampshire. But the left hemisphere also contains regions that specialize exclusively in language and the right has some regions that exclusively serve musical perception, he said. In the brains of musical idiot savants—individuals who are talented musicians despite severe mental retardation—the dichotomy is especially pronounced.

Researchers are beginning to model these special music circuits in computer neural networks and to map them in living brains, Dr. Bharucha said. Some of the biology is known. When sound waves enter the human ear, they stimulate neurons called "hair cells" that lie on a flat plane. Depending where it is located, each hair cell responds to a characteristic frequency. Those at one end respond to high frequencies, those at the other end to low frequencies.

The signals are then passed up through the brain stem, where information from both ears is integrated to help locate the origin of the sounds. From there they enter the primary auditory cortex, which contains cells that specialize in particular frequencies. Simple sounds are again mapped on neural tissue.

At these lower levels, language and music have shared pathways, Dr. Bharucha said. The question is, at what point do they diverge? Where are the neural circuits for music and which qualities of music do they specialize in recognizing?

"We predict you should find cells that are tuned to familiar chords," Dr. Bharucha said. Some cells may be tuned for octaves, musical fifths and fourths, he said. Others would specialize in detecting patterns of ascending or descending tones.

There even may be cells that are wired together to encode familiar songs and melodies, Dr. Bharucha said. The more familiar the song, the fewer the neurons needed for this task. The idea is supported by classical experiments in which doctors inserted tiny electrodes into the brains of awake patients. As the active electrode was moved from one tiny clump of cells to another, patients reported hearing different songs, fragments of symphonies or familiar voices.

Studies of brain-damaged patients shed considerable light on these circuits, said Dr. Isabelle Peretz, a psychologist at the University of Montreal. She has studied three patients who suffered damage to both the left and right temporal lobes—where major auditory circuits reside—and as a result have lost the ability to recognize familiar songs, the condition called amusia. Only music is affected, Dr. Peretz said. These patients can recognize human voices, animal cries, traffic sounds and all other auditory information with no trouble. The patients say they can still respond emotionally to music, even though the neural network that recognizes melody and other musical qualities has been destroyed.

Clues to what has been lost may be found in connections between the temporal lobes and frontal lobes, where higher-order decisions are made, said Dr. Robert Zatorre, a psychologist at the Montreal Neurological Institute and Hospital. "Music evolves over time and to decode it you need to hold information in working memory," he said. "You need a sophisticated buffer to relate an event happening now to an event twelve seconds ago. We think this is happening in the frontal and temporal lobe interaction. The frontal lobe runs the traffic here, holding the relevant information." Depending on what you are listening for in a particular piece of music, the relevant information may be repeated themes, for instance, or timbres of different instruments.

Music can also be imagined, as Dr. Zatorre points out, because people have stored representations of songs, melodies and the sounds of instruments. These representations can be stimulated internally, he said. When a song is "imagined," the cells and networks that are activated are identical to those used when a person actually hears music coming from the external world. But when songs are imagined, Dr. Zatorre said, parts of the visual cortex also light up, suggesting that tonal patterns evoke visual imagery patterns.

"We don't know what triggers musical imagery," Dr. Zatorre said, "but it is very common for people to wake up in the morning with songs running through their heads."

Brain mapping shows that musical networks also extend into the brain's emotional circuits in the limbic system, Dr. Zatorre said. People report strong emotional sensations when they listen to music, saying, for example, that they feel like their hair is standing on end or that they have a lump in their throat, he said. How these circuits are wired up has yet to be determined. Dr. Fernando Nottebohm, a neuroscientist at Rockefeller University in New York, suggests that bird songs may also have an emotional

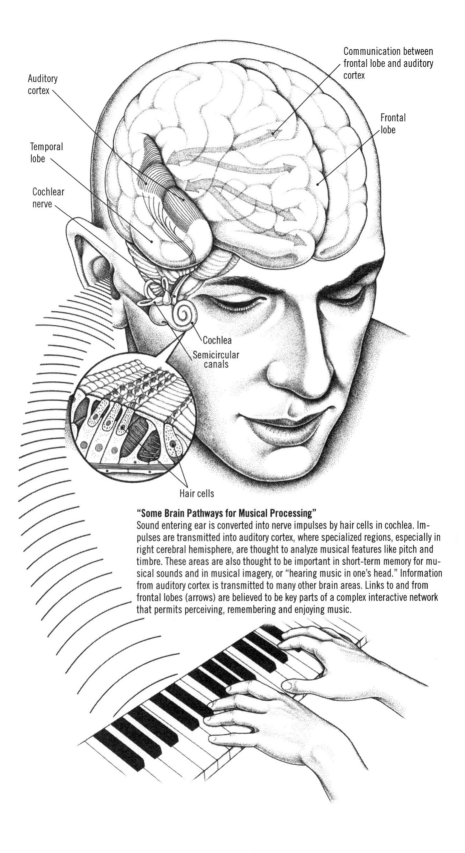

"Some Brain Pathways for Musical Processing"
Sound entering ear is converted into nerve impulses by hair cells in cochlea. Impulses are transmitted into auditory cortex, where specialized regions, especially in right cerebral hemisphere, are thought to analyze musical features like pitch and timbre. These areas are also thought to be important in short-term memory for musical sounds and in musical imagery, or "hearing music in one's head." Information from auditory cortex is transmitted to many other brain areas. Links to and from frontal lobes (arrows) are believed to be key parts of a complex interactive network that permits perceiving, remembering and enjoying music.

component. "I believe some birds sing purely for pleasure," he said. "I wouldn't be surprised if a male songbird singing on top of a tree is having a glorious time."

Research into how the brain decodes music reveals insight about what makes music so interesting to most people. It is a matter of surprise, researchers say. The brain becomes accustomed to patterns of music based on exposure to various musical traditions, Dr. Bharucha said. Groups of cells may be especially attuned to sounds heard in rock and roll, Mozart or riffs on a sitar. When the brain hears, say, 10 notes of a melody, it will predict the 11th note based on these stored connections. When the note is predicted correctly, he said, the cellular connections become even stronger. If a note is slightly off, it can be either jarring or aesthetically pleasing.

It is the violation of these brain-based expectations that makes music interesting, Dr. Bharucha said. Composers regularly exploit them. For example, the note-to-note violations in classical music are very subtle. "At the beginning of a Handel violin sonata, the violin plays notes that lead you to expect completion of a chord with a D on the musical scale," he said. "Instead, you hear an E, which is a subtle violation in the same key in the same scale.

"If you want more serious violations, you pick notes out of key," Dr. Bharucha continued. "Western music pushed the limit of these violations until, in the early twentieth century, the whole thing collapsed." Composers like Schoenberg minimized the degree of expectation completely, but it never caught on with popular audiences. Minimalist music by composers such as Philip Glass is a revenge against excess surprise, he said. It is very predictable, some would say boring.

Rock and roll, Dr. Bharucha said, is often thought to be a new form of music, but in fact it uses chords and elemental patterns that go back centuries. Ironically, as far as pitches and harmonies go, rock and roll is more traditional and has changed less than classical music, he said, which may help explain its popularity.

Musical talent is another mystery. As with any kind of intelligence, the neural maps that serve music perception may be stronger or larger or better in some people. Researchers have reported that people with perfect pitch have in their left hemispheres highly developed structures associated with musical perception. But there is no single musical talent, said Dr. Peter Ostwald, a psychiatrist at the University of California School of Medicine in San Francisco. There may be separate talents for tone recognition, melodic structure, movement, ability to play an instrument well and the gift for dramatizing oneself and playing in public.

Nevertheless, early exposure to music and musical training does make a difference in musical talent. Like language, music follows a course in infant and child development, Dr. Zatorre said. Six-month-old babies are sensitive to musical patterns. A two-year-old hearing the *Barney* song will pick it up and start singing it incessantly. Children sing songs to themselves as they play.

At the University of California at Irvine, researchers built computer models of how cells in the auditory cortex might fire together during learning. When the computer program was connected to a device that translated its mathematical code into sounds, musical themes appeared.

"It got us to thinking," said Dr. Gordon Shaw, a physicist at Irvine's Center for the Neurobiology of Learning and Memory. He said that the way cells are connected throughout the cortex might compose the basic neural language of the brain. "When you hear music, you are exciting inherent brain patterns that derive from this structure and connectivity.

"Then we made another big jump," he added. "Musical training at an early age might reinforce these patterns. Music is structured in space and time. Could it enhance or strengthen the circuits that help you think and reason in space and time?"

To find out, Frances Rauscher, an Irvine researcher, has been working with preschool children in Los Angeles. One group of three-year-olds received weekly piano lessons and participated in daily sessions of group singing. Another group did not get the extra training. After a year, she said, the musically trained children scored 80 percent higher on tests of spatial and temporal reasoning, an ability that underlies many kinds of mathematics and engineering.

Could this explain why so many physicists and mathematicians are also gifted musicians? Scientists who study music and the brain may soon find the answer.

The first article in this series, about evidence for a new theory of consciousness, appeared on March 21. The second article, about the way the brain processes emotion, appeared on March 28. The third article, about the brain's "working memory," appeared on May 2.

—Sandra Blakeslee, May 1995

Odd Disorder of Brain May Offer New Clues

AT AGE 18, WENDY VEROUGSTRAETE had high ambitions. "You are looking at a professional book writer," she said cheerfully. "My books will be filled with drama, action and excitement. And everyone will want to read them. I am going to write books, page after page, stack after stack. I'm going to start on Monday."

But that Monday never came. Although she composes lyrics to love songs, has a rich vocabulary and tells wondrous stories, Ms. Verougstraete has an I.Q. of only 49. She cannot tie her shoes, set a table, cross the street alone or make change for a quarter. Her reading, writing and drawing skills are like those of a first grader. Now age 25, she lives in a group home for mentally retarded adults.

Ms. Verougstraete has Williams syndrome, an enigmatic birth disorder caused by the loss of one copy of the gene that makes elastin, a protein that is the chief constituent of the body's elastic fibers, and possibly by the loss of another gene or genes of unknown function that lie next to elastin on chromosome 7.

The result of this small genetic loss is far-reaching. There are severe malformations throughout the brain and heart, yet the capacity for language is remarkably unaffected. If anything, language and sociability are enriched. Williams syndrome children, who have distinctive elfin features, are extremely social, verbal and adept at recognizing faces, but most cannot expect to live independent lives.

Cognitive neuroscientists say that Williams syndrome, first described in 1961, presents an unparalleled opportunity to probe the deepest mysteries of the human brain. What are the genetic origins of language and sociability? What do we mean by intelligence? Which genes determine the brain's basic architecture, controlling how it is wired during fetal development? How does a young child's brain compensate for inborn deficits by rewiring itself in alternative circuits? And how do genes contribute to complex behaviors such as personality?

Williams syndrome may also help resolve the huge debate in cognitive psychology over the nature of language, said Dr. Albert Galaburda, a neurologist at Harvard University Medical School. "Is it special from the word go, under the control of special genes and located in special parts of the brain?" he said. "Or does it piggyback on general mental function and intelligence? Williams children suggest language is unique because there is a genetic defect that spares it."

At a meeting in La Jolla, California, the world's leading experts on Williams syndrome presented their latest explorations of the biological links between genes and behavior. Efforts to forge such links in studies of other behaviorally complex disorders, such as schizophrenia and manic depression, have not been very successful. But Williams syndrome researchers say that they have an advantage in knowing the exact locus of a genetic defect that results in a remarkably consistent behavioral profile. The meeting was sponsored by the Williams Syndrome Association and the Salk Institute for Biological Studies.

Dr. Ursula Bellugi, director of the Laboratory for Cognitive Neurosciences at the Salk Institute, has carried out extensive studies of Williams syndrome children. Her interest began several years ago after a late-night phone call from Nancy Verougstraete, Wendy's mother, who had just read a magazine article by Noam Chomsky on the biological basis of language. "I want you to meet my daughter," Mrs. Verougstraete said. "She's retarded but has good language. I think you should investigate her unusual abilities."

"The child, who was thirteen, came in and I was puzzled," Dr. Bellugi said in an interview. "She had a very unusual profile. Her grammar was complex and without error. Her word use was rich, but general cognition and problem solving were very impaired. She had been placed in a school for the mentally retarded but her teachers did not know how to deal with her. I looked around for information on cognitive abilities of Williams syndrome but very little was known."

Williams syndrome occurs in one of every 20,000 births. Many of the children have elevated levels of calcium in their blood during infancy, which is thought to make them extremely cranky. Parents say their babies cry nonstop for several months. All have heart defects, typically a narrowing of the aorta or pulmonary arteries.

Williams children have similar faces, with an upturned nose, wide mouth, full lips, small chin and puffiness around the eyes. Those with blue or green eyes have a prominent starburst pattern on their irises. Their voices are hoarse. All are mentally retarded on standard I.Q. tests but to different degrees. Although some can attend regular classes, most require special education.

Williams children are typically late in every aspect of development, Dr. Bellugi said, including language. But when grammar develops, often around age four, she said, it takes off with remarkable speed.

To probe the mental peaks and valleys of Williams syndrome, Dr. Bellugi and her colleagues abandoned standardized I.Q. tests and developed a battery of experimental probes for specific domains of intelligence. For the sake of comparison, she gave the same tests to children with Down syndrome matched for age, sex and I.Q.

In general problem solving, children with Williams and Down syndrome are very similar, she said. But in linguistic probes, the groups are very different.

On vocabulary tests, Williams children display a predilection for unusual words, Dr. Bellugi said. Asked to name as many animals as they can think of in one minute, they come up with creatures such as ibex, Chihauhua, saber-toothed tiger, weasel, crane and newt. Children with Down syndrome give simple examples like dog, cat and mouse, she said.

When Williams syndrome children tell stories, they make their voices come alive with drama and emotion, Dr. Bellugi said. They set the mood by whispering certain lines or punctuating dialogue with audience attention-getters such as "Gadzooks" or "lo and behold!" Children with Down syndrome, on the other hand, tell simple stories without emotion.

Williams children employ a rich variety of grammatically complex forms, including passive sentences, conditional clauses and embedded relative clauses, Dr. Bellugi said. They can correct ungrammatical sentences.

One of the most salient feature of Williams children is their extreme sociability, she said. They seem drawn to adults and will carry on long conversations, asking endless questions and making flattering comments.

Parents say that many Williams children have excellent musical ability and enjoy playing instruments and singing songs, she said. Moreover, they tend to be sensitive to certain classes of sounds, some of which are disturbing while others turn into obsessions. One boy loves vacuum cleaners. Another adores power lawn mowers. And these children's hearing is acute; they can detect faint sounds in the distance, long before others notice them.

Dr. Helen Neville, then a professor of cognitive neuroscience at the University of California at San Diego, has examined the brain waves of Williams children in response to different auditory stimuli. While their hearing tests are in the normal range, neurons in the auditory cortex are exceptionally excitable, Dr. Neville said. These cells respond to tones in a characteristic manner, she said, and may one day be used in the diagnosis of Williams syndrome.

"It's amazing when you look at kids who can't tie shoes or figure out many simple tasks, yet they talk like angels," Dr. Bellugi said. "They can explain all sorts of things but can't do them. They know a lot about words but have very little world knowledge."

Animate and inanimate objects have distinct traits, yet "a fourteen-year-old Williams child will insist that the moon is alive because it moves," Dr. Bellugi said. "It's not that they don't have categories of knowledge, but they don't reorganize their understanding of knowledge in ways that normal children do as they get older."

Their spatial abilities are particularly impoverished, Dr. Bellugi said. They cannot draw simple objects, nor can they match two parallel lines in a field of oblique lines. While some Williams children can gain rudimentary reading and writing skills, most are never able to do simple arithmetic.

The genetic defect that leads to Williams syndrome was discovered in 1993 by researchers at the University of Utah in Salt Lake City. Medical geneticists were studying an inherited heart defect, super valvular aortic stenosis, and found that it is caused by partial deletions of one elastin gene on chromosome 7.

Although children with this heart defect are not mentally retarded as a result, all of those with Williams syndrome have the heart defect, said Dr. Colleen Morris, who worked on the Utah team and is now an associate professor at the University of Nevada School of Medicine in Las Vegas.

When researchers looked at chromosome 7 in Williams children, they found that every child was missing a full elastin gene plus a length of DNA surrounding it. The neighboring DNA is known to contain one or more genes, but they have not yet been identified. Unlike the heart defect alone, Dr. Morris said, Williams syndrome occurs sporadically. It does not run in families.

Researchers believe that the brain abnormalities in Williams syndrome are caused by this genetic deletion. Elastin and the mystery gene or genes next to it could play a key regulatory role in brain development, said Dr. Galaburda, who has done two autopsies on the brains of two young people with Williams syndrome. But how the genes determine morphology is unknown.

Dr. Bellugi and her colleagues at first thought that the brains of Williams children might be asymmetrical. After all, they had good language ability, which is controlled predominantly in the left side of the brain, and poor spatial ability, which tends to originate in the right side. But, in one of the biggest mysteries of Williams syndrome, the children are extremely adept at identifying faces—a difficult spatial task carried out by the right hemisphere.

But Dr. Galaburda's autopsies and numerous brain-imaging studies present a more complicated picture. There are no obvious lesions in the left or right sides of the brain, she said. Instead, the cerebral cortex is shrunken on both sides and fibers connecting the left and right hemispheres are thin or diminished. Moreover, these abnormally small brain regions are packed with an excessive number of neurons.

During development, the brain produces an overabundance of neurons and then eliminates the excess through programmed cell death. "My hunch is that there may be diminished cell death" in Williams brains, Dr. Galaburda said.

At the same time, one circuit in the Williams brain seems spared. The frontal lobes, medial temporal lobes that include part of the auditory cortex and the neocerebellum are closer to normal size, Dr. Galaburda said.

This is an utterly fascinating finding, said Dr. Eric Courchesne, an expert on autism at the Children's Hospital and Health Center in San Diego. In evolutionary terms, the cerebellum is an old part of the brain involved in movement. It is lavishly connected to most other parts of the brain and "allows you to rapidly and smoothly change your attention," Dr. Courchesne said.

The neocerebellum, a thin layer of cells atop the older brain region, evolved more recently along with the prefrontal cortex, Dr. Courchesne said. Humans are the only creatures with large versions of these two brain regions and the only ones with language. Moreover, brain-imaging studies show that the neocerebellum is activated only when semantic reasoning is required.

Could this be a recently evolved circuit underlying language and sociability, one that is spared in Williams syndrome? Does the gene deletion that marks the syndrome impair an older set of instructions for brain development, while leaving a newer language circuit intact?

These are just speculations, Dr. Courchesne said. But they are intriguing. The neocerebellums of autistic children—who are antisocial and poor at language, but good at spatial tasks—are tiny compared to normal and Williams children.

Finally, the Williams children provide an excellent opportunity to study domains of intelligence. I.Q. is a gross measure that does not capture the strengths and weaknesses of each individual, Dr. Bellugi said, as Williams children demonstrate so dramatically.

—Sandra Blakeslee, August 1994

Brain May Have Separate Units to Digest Writing and Speech

THE HUMAN BRAIN HAS SEPARATE systems for processing written and spoken language, a study suggests.

The research, described in the journal *Nature,* supports a growing body of evidence that the brain divides complex tasks into compartments and assigns specific functions to independent subsystems.

The phenomenon has been shown in tasks like vision, but this is the first time it has been demonstrated in language, according to Dr. Alfonso Caramazza, professor and director of cognitive science at the Johns Hopkins University in Baltimore, who wrote the study.

Dr. Caramazza studied two stroke patients who had trouble with verbs. One made errors in oral tasks and the other made errors in written tasks. In an experiment, Dr. Caramazza demonstrated that the patients stored and processed verbs in separate systems, one for written language and one for spoken language.

"This is a fascinating finding," said Dr. Steven Pinker, a professor in the department of brain and cognitive science at the Massachusetts Institute of Technology. It touches, he said, on one of the hottest questions in cognitive neuroscience: is the brain a huge network of interconnections that gains complexity by following simple rules of learning, or is it a network of dedicated compartments that carry out tasks independently?

Those who believe in simple rules of learning say that only one system of grammar can exist in the brain, said Dr. Pinker. He said this theory holds that when people speak or write or think about language, they draw on this single source to carry out all language tasks.

Dr. Caramazza supports the compartmentalization school of thought. "You can think of the brain as you think of a factory," he said in a telephone interview. "When you build a car, someone works on the wheels, the pistons, the frame and so on. Each group has a specific function to carry out."

Similarly, the brain assigns tasks to specific neural subsystems in carrying out complex tasks, of which vision is a good example, he said. When the brain "sees" an object, Dr. Caramazza said, separate subsystems process edges, color, contour, shades of gray, motion and other properties of three-dimensional space.

Language is also broken down into components, Dr. Caramazza said. When the brain processes words, it uses separate neuronal subsystems to handle sounds of words, written forms of words, word roots, affixes and other properties of language.

To demonstrate this phenomenon, Dr. Caramazza turned to brain-damaged patients. "We've known for some time that patients can have selective damage that affects either nouns or verbs," he said. "But previous studies have not allowed us to be precise about the deficit. Where is the grammatic information represented? Is there a single grammar or is grammar mapped onto spoken and written forms of language?"

A patient called H.W. made mistakes only in speaking, he said, while a patient called S.J.D. made mistakes only in writing. Both had difficulty with verbs.

In the experiment, Dr. Caramazza presented both patients with homonyms, words that sound the same and are spelled the same but have different meanings and are different parts of speech. "Walk" can mean a walk in the park or to walk down the street, he said, but there is nothing about the word standing alone that indicates whether it is a noun or a verb. The brain must make the distinction at some abstract level of processing grammar.

Both patients were presented with two sentences—"There is a crack in the wall" and "Don't crack nuts in here"—in which the word "crack" was underlined, said Dr. Caramazza. They were asked to read the sentences silently and then to say the underlined word out loud. Next, they were given the same sentences in written form but the word crack was omitted. The sentences were then read aloud to them and they had to fill in the blank with the word "crack."

H.W. could do the writing task and could speak the word "crack" when used as a noun, Dr. Caramazza said, but she could not say the word "crack" as a verb. She was at a loss for words in that situation.

S.J.D. could do the speaking tasks and could write "crack" as a noun, he said, but she could not write the word when it was presented to her as a verb.

This suggests that verbs are stored in two different lexicons, Dr. Caramazza said, one for speaking and one for writing. "We have a stock of knowledge in written form and a stock of knowledge in spoken form," he said.

Adjectives and nouns are probably processed similarly, he said, although further experiments will be needed to prove it. Dr. Caramazza said he would like to perform the same experiment on stroke patients who have selectively lost the ability to discriminate and use nouns.

—SANDRA BLAKESLEE, February 1991

Subtle But Intriguing Differences Found in the Brain Anatomy of Men and Women

RESEARCHERS WHO STUDY THE BRAIN have discovered that it differs anatomically in men and women in ways that may underlie differences in mental abilities.

The findings, although based on small-scale studies and still very preliminary, are potentially of great significance. If there are subtle differences in anatomical structure between men's and women's brains, it would help explain why women recover more quickly and more often from certain kinds of brain damage than do men, and perhaps help guide treatment.

The findings could also aid scientists in understanding why more boys than girls have problems like dyslexia, and why women on average have superior verbal abilities to men. Researchers have not yet found anything to explain the tendency of men to do better on tasks involving spatial relationships.

These findings are emerging from the growing field of the neuropsychology of sex differences. Specialists in the discipline met at New York University Medical Center to present their latest data.

Research on sex differences in the brain has been a controversial topic, almost taboo for a time. Some feminists fear that any differences in brain structure found might be used against women by those who would cite the difference to explain "deficiencies" that are actually due to social bias. And some researchers argue that differences in mental abilities are simply due to environmental influences, such as girls being discouraged from taking math seriously.

The research is producing a complex picture of the brain in which differences in anatomical structure seem to lead to advantages in performance on certain mental tasks. The researchers emphasize, however, that it is not at all clear that education or experience do not override what differences in brain structure contribute to the normal variation in abilities. Moreover, they note that the brains of men and women are far more similar than different.

Still, in the most significant findings, researchers are reporting that parts of the corpus callosum, the fibers that connect the left and right hemispheres of the brain, are larger in women than men. The finding is surprising because, over all, male brains—including the corpus callosum as a whole—are larger than those of females, presumably because men tend to be bigger on average than women.

Because the corpus callosum ties together so many parts of the brain, a difference there suggests far more widespread disparities between men and women in the anatomical structure of other parts of the brain.

"This anatomical difference is probably just the tip of the iceberg," said Sandra Witelson, a neuropsychologist at McMaster University Medical School in Hamilton, Ontario, who did the study. "It probably reflects differences in many parts of the brain which we have not yet even gotten a glimpse of. The anatomy of men's and women's brains may be far more different than we suspect."

The part of the brain which Dr. Witelson discovered is larger in women is in the isthmus, a narrow part of the callosum toward the back. Her findings were published in the journal *Brain*.

Dr. Witelson's findings on the isthmus are based on studies of 50 brains, 15 male and 35 female. The brains examined were of patients who had been given routine neuropsychological tests before they died.

"Witelson's findings are potentially quite important, but it's not clear what they mean," said Bruce McEwen, a neuroscientist at Rockefeller University. "In the brain, bigger doesn't always mean better."

In 1982 a different area of the corpus callosum, the splenium, was reported by researchers to be larger in women than in men. But that study was based on only 14 brains, five of which were female. Since then, some researchers, including Dr. Witelson, have failed to find the reported difference, while others have.

Since such differences in brain structure can be subtle and vary greatly from person to person, it can take the close examination of hundreds of brains before neuroanatomists are convinced. But other neuroscientists say the findings are convincing enough to encourage them to do tests of their own.

Both the splenium and the isthmus are located toward the rear of the corpus callosum. This part of the corpus callosum ties together the cortical areas on each side of the brain that control some aspects of speech, such as the comprehension of spoken language and the perception of spatial relationships.

"The isthmus connects the verbal and spatial centers on the right and left hemispheres, sending information both ways—it's a two-way highway," Dr. Witelson said. The larger isthmus in women is thought to be related to women's superiority on some tests of verbal intelligence. It is unclear what, if anything, the isthmus might have to do with the advantage of men on tests of spatial relations.

The small differences in abilities between the sexes have long puzzled researchers.

On examinations like the Scholastic Aptitude Test, which measures overall verbal and mathematical abilities, sex differences in scores have been declining. But for certain specific abilities, the sex differences are still notable, researchers say.

While these differences are still the subject of intense controversy, most researchers agree that women generally show advantages over men in certain verbal abilities. For instance, on average, girls begin to speak earlier than boys and women are more fluent with words than men, and make fewer mistakes in grammar and pronunciation.

On the other hand, men, on average, tend to be better than women on certain spatial tasks, such as drawing maps of places they have been and rotating imagined geometric images in their minds' eye—a skill useful in mathematics, engineering and architecture.

Of course, the advantages for each sex are only on average. There are individual men who do as well as the best women on verbal tests, and women who do as well as the best men on spatial tasks.

One of the first studies that directly links the relatively larger parts of women's corpus callosums to superior verbal abilities was reported at the joint meeting of the New York Neuropsychology Group and the New York Academy of Sciences by Melissa Hines, a neuropsychologist at the University of California at Los Angeles Medical School.

Dr. Hines and her associates used magnetic resonance imaging, a method that uses electrical fields generated by the brain, to measure the brain anatomy of 29 women. They found that the larger the splenium in the women, the better they were on tests of verbal fluency.

There was no relationship, however, between the size of their splenium and their scores on tests of spatial abilities, suggesting that differences in those abilities are related to anatomical structures in some other part of the brain or have nothing to do with anatomy.

"The size of the splenium," Dr. Hines said, "may provide an anatomical basis for increased communication between the hemispheres, and perhaps as a consequence, increased language abilities."

Researchers now speculate that the larger portions of the corpus callosum in women may allow for stronger connections between the parts of women's brains that are involved in speech than is true for men.

"Although we are not sure what a bigger overall isthmus means in terms of microscopic brain structure, it does suggest greater interhemispheric communication in women," Dr. Witelson said. "But if it does have something to do with the cognitive differences between the sexes, it will certainly turn out to be a complex story."

Part of that complexity has to do with explaining why, despite the bigger isthmus, women tend to do less well than men in spatial abilities, even though the isthmus connects the brain's spatial centers, too.

"Bigger isn't necessarily better, but it certainly means that it's different," Dr. Witelson said.

A variety of other differences in the brain have been detected by the researchers in their studies.

For instance, Dr. Witelson found in her study that left-handed men had a bigger isthmus than did right-handed men. For women, though, there was no relationship between hand preference and isthmus size.

"How our brains do the same thing, namely use the right hand, may differ between the sexes," Dr. Witelson said.

She also found that the overall size of the callosum, particularly the front part, decreases in size between 40 and 70 years of age in men, but remains the same in women.

Several converging lines of evidence from other studies suggest that the brain centers for language are more centralized in men than in women.

One study involved cerebral blood flow, which was measured while men and women listened to words that earphones directed to one ear or the other. The research, conducted by Cecile Naylor, a neuropsychologist at Bowman Gray School of Medicine in Winston-Salem, North Carolina, showed that the speech centers in women's brains were connected to more areas both within and between each hemisphere.

This puts men at a relative disadvantage in recovering from certain kinds of brain damage, such as strokes, when they cause lesions in the speech centers on the left side of the brain. Women with similar lesions, by contrast, are better able to recover speech abilities, perhaps because stronger connections between the hemispheres allow them to compensate more readily for damage on the left side of the brain by relying on similar speech centers on the right.

In the *Journal of Neuroscience,* Roger Gorski, a neuroscientist at UCLA, reported finding that parts of the hypothalamus are significantly bigger in male rats than in female ones, even though the size of the overall brain is the same in both sexes.

And Dr. McEwen, working with colleagues at Rockefeller University, has found a sex difference in the structure of neurons in part of the hippocampus that relays messages from areas of the cortex.

Dr. McEwen, working with rats' brains, found that females have more branches on their dendrites, which receive chemical messages to other neurons, than do males. Males, on the other hand, have more spines on their dendrites, which also receive messages from other neurons. These differences in structure may mean differing patterns of electrical activity during brain function, he said.

"We were surprised to find any difference at all, and, frankly, don't understand the implications for differences in brain function," Dr. McEwen said. "But we'd expect to find the same differences in humans; across the board, findings in rodents have had corollaries in the human brain."

In the corpus callosum, millions of fibers link many parts of the brain, including a region in the left hemisphere involved in speech and a region in the right hemisphere involved in spatial perception. One study has found that the isthmus, a narrow part of the corpus callosum toward the back of the head, is larger in women than in men. Another area, the splenium, was found to be larger in women in some studies.

—Daniel Goleman, April 1989

Men and Women Use Brain Differently, Study Discovers

USING A POWERFUL METHOD for glimpsing the brain in action, researchers have found the first definitive evidence that men and women use their brains differently.

The investigators, who were seeking the basis of reading disorders, asked what areas of the brain were used by normal readers in the first step in the process of sounding out words. To their astonishment, they discovered that men use a minute area in the left side of the brain while women use areas in both sides of the brain.

Dr. Sally E. Shaywitz, a behavioral scientist at the Yale University School of Medicine who was a principal author of the study, said, "As far as I know, this is the first time that anyone has been able to demonstrate anything functionally different" between the brains of men and women.

The findings follow on a rich body of research looking for sexual differences in the brain. Psychologists have found that women do better on certain tests, like those measuring verbal speed, and that men do better on other tasks, like imagining what an object would look like if it were rotated. Neurologists have found that women seem to recover better from strokes in the left hemisphere of the brain, where language abilities are thought to be situated. Autopsy studies have shown that male brains are more asymmetrical than female brains.

But these previous studies were indirect. The psychological studies could not prove that it was nature, not nurture, that elicited the differences. The anatomical studies could not show what the actual effects of the brain differences were.

The study, in contrast, showed actual differences in the parts of the brain used when men and women were thinking, and coming up with the same answers.

Dr. Shaywitz said the finding meant that "the brain is a lot more complicated than people envisioned." But it does not say that women's brains are better at this task than men's or vice versa. Although the men and women used their brains differently, she added, the fact that they sounded out words equally well means that "the brain has a lot of different ways to get to the same result."

Dr. Elizabeth Hampson, a neuroscientist at the University of Western Ontario, said the finding "provides definitive evidence" that men and women can use their brains differently to perform the same task. "Nothing was conclusive until now," she said. It means, she said, "we should be open to that possibility in other areas of the brain as well."

Dr. Shaywitz said she was particularly surprised to see differences between men and women in decoding words. Reading, she said, has nothing to do with basic survival skills or reproduction, for which men and women might have developed different brain functions during the long course of evolution. "This is a difference that involves cognition," she said. "And it is the most complex of human functions. Reading is the pinnacle of what humans can do."

Dr. Reid Lyon, director of extramural research on language disorders and learning disabilities at the National Institute of Child Health and Human Development, said the discovery went far beyond a mere curiosity. It is a huge step in a comprehensive research program that is allowing scientists to understand why some children and adults have such difficulty learning to read and it has immediate implications for tests for reading disabilities and strategies to overcome them, he said.

It also might explain why it is that although girls and boys are equally likely to have difficulty learning to read, women seem to compensate better than men for their disability, he said.

Dr. Shaywitz, her husband, Dr. Bennett A. Shaywitz, a neurologist, and their colleagues published their findings in the journal *Nature*.

The Shaywitzes' main interest has been to understand the neurological underpinnings of reading disorders. In recent years, they and others have discovered that dyslexia, a difficulty in learning to read, is not a visual problem, not a problem that involves reversing letters, but is instead a language problem. People with reading disabilities have great difficulty breaking words into their most basic units, phonemes. A test for an ability to separate words into phonemes might be to ask a person to say philosophy without the "la" sound or Germany without the "ma."

Dr. Bennett Shaywitz explained that "by convention, we have adopted rules that say that certain sounds in print correspond to certain letters and certain letters form phonemes. The trick in reading is to get from print into phonemes, then it's automatic."

In studying people with reading disabilities who nonetheless learned to read, Dr. Sally Shaywitz said she and others found that they learned to compensate for their inability to decode words into phonemes. "They used other higher order cognitive abilities: their vocabulary, their reasoning power," she explained. But, she said, reading was never easy or automatic for these people.

Researchers have discovered a method that can potentially show the brain in action and that is completely harmless to subjects.

The method, functional magnetic resonance imaging, can show differences between blood carrying oxygen and blood that is depleted of oxygen. An active area of the brain gets fresh supplies of oxygenated blood. The trick to using functional MRI's is to design thinking experiments carefully so that they will isolate a particular task.

The Shaywitzes and their colleagues asked their subjects, 19 men and 19 women, to recognize whether nonsense words, like "jete" and "lete," rhyme. The only way to know is to sound them out. And the only way to do that is to use phonemes.

All 19 men used a tiny area of the brain, about a centimeter, or less than half an inch, in diameter, near Broca's region, an area in the front of the left side of the brain, at the temple, that is thought to be used in producing speech. Eleven of the 19 women used this region but also used a comparable area on the right side of the brain. The rest of the women used just the left side of the brain.

The next step, now under way, is to see how children and adults with reading disorders use their brains. The Shaywitzes, who will not comment yet on what they are finding, hinted that those with reading disabilities had profoundly different functional MRI patterns.

Soon, Dr. Lyon said, it should be possible to design simple tests that will reliably detect young children who will have reading disabilities before they are faced with the daunting task of learning to read. It may also be possible to design ways to help these children overcome or even avoid the coming struggle.

Shirley Kramer, executive director of the National Center for Learning Disabilities, an advocacy group based in New York, said she could hardly wait. The work, she said, "has moved this field forward by leaps and bounds," adding, "There is a real momentum going now."

—Gina Kolata, February 1995

5

LANGUAGE AND SOCIETY

Though language is spoken or written, words are only part of it. Gestures, silences, the context that imbues a single phrase with a whole scene of meaning—these are the other half of the communications system. Foremost among the nonverbal elements of language is laughter.

Over the centuries, laughter and gesture tend to remain more constant than the words. Languages are in such a rapid state of flux that even the most widely spoken seem destined to have limited lives. Even English, the first or second language in many continents and embedded as the international language of science, may not last forever, at least in its present form. Shakespeare often sounds archaic to modern ears; Chaucer is halfway to being a foreign language. The English of today will surely sound quaint to readers three centuries from now.

Most extinct languages are lost in perpetuity, but a few survive in whispers of their former selves. These are the languages that fell extinct since the invention of writing. Latin and ancient Greek are dead but not extinct, since they survive as liturgical languages and through copious written literatures.

Other languages survive only as texts, preserved in such durable materials as stone or baked clay. Linguists have had surprising success in translating these mysterious writings, although it has often taken decades for inspired insight to seize upon some subtly hidden clue that yields the secrets of the rest of the text.

But with the remarkable exception of Hebrew, resurrected as the living language of Israel, the road to linguistic extinction is one way. As indigenous peoples are coopted into the larger modern cultures around them, living languages are fast disappearing. Interviewing the last elderly speakers, anthropologists are trying to record as much as possible before the final silence.

Laughs: Rhythmic Bursts of Social Glue

HERE IS A SAMPLING OF knee slappers to jump-start your day:
"Got to go now!"
"I see your point."
"It must be nice."
"Look! It's Andre."

Hey, wait a minute. Where are the guffaws, the chuckles, at the very least a polite titter or two? Get me laugh track! Doesn't this deadbeat crowd know that such lines are genuine howlers, field-tested fomenters of laughter among ordinary groups of people in ordinary social settings?

We're not talking Aristophanes here, or even Phyllis Diller. We're talking the sort of laughter that we give and receive every day while strolling with friends in the park or having lunch in the company cafeteria or chatting over the telephone. The sort of social laughter that punctuates casual conversations so regularly and unremarkably that we never think about or notice it—but that we would surely, sorely miss if it were gone.

One person who has thought about and noticed laughter in great detail is Dr. Robert R. Provine, a professor of neurobiology and psychology at the University of Maryland Baltimore County. Dr. Provine has become a professional laugh tracker, if you will, an anthropologist of our amusement, asking the deliciously obvious questions that science has not deigned to consider before. He has analyzed what, physically, a laugh is, what its vocal signature looks like and how it differs from the auditory shape of a spoken word or a cry or any other human utterance.

He has asked when people laugh and why, what sort of comments elicit laughter, whether women laugh more than men, whether a person laughs more while speaking or while listening. He has studied the rules of laughter: when in a conversation a laugh will occur, and when, for one reason or another, the brain decides it is taboo. He has compared human laughter to the breathy, panting vocalizations that chimpanzees make while they are being chased or tickled, and that any primatologist or caretaker will firmly describe as chimpanzee laughter.

Dr. Provine has eavesdropped on 1,200 bouts of laughter among people in malls and other public places, noting down the comments that preceded each laugh and compiling a list of what he calls his "greatest hits" of laugh generators, which include witlessisms like those quoted above. In so doing he has made a discovery at once startling and perfectly sensible: most of what we laugh at in life is not particularly funny or clever but merely the stuff of social banter, the glue that binds a group together. Even the comparatively humorous laughgetters are not exactly up to Seinfeld, lines like, "She's got a sex disorder: she doesn't like sex"; or "You don't have to drink. Just buy us drinks." It is probably a good thing that our laughmeters are set so low, because very few of us are natural wits, and those that are often get into bad moods and refuse to say a single clever thing for entire evenings at a time.

Dr. Provine, a tall man with a well-groomed academic-issue beard who in profile looks faintly like the actor Fernando Rey, is neither clownish nor severe, somehow remaining animated about his subject without becoming silly. He can laugh loudly on command to demonstrate his points, which is something many people refuse to do. In videos, Dr. Provine is shown approaching strangers on the Baltimore waterfront, telling them he is studying laughter and asking them to laugh for him. Usually, people give sidelong glances to their companions, grin, fuss with their hair, and as he persists, they grow annoyed. "I can't laugh on command," they complain. "Tell me a joke first."

To Dr. Provine, that difficulty reveals something important about the nature of laughter. We can smile on command, albeit stiffly, and we can certainly talk on command, but laughter has an essential spontaneous element to it. It is a vocalization of a mood state, rather than a cognitive act, and as such it is difficult to fake, just as it is hard to force out tears. Those who are good at laughing on cue, said Dr. Provine, often have stage experience.

Dr. Provine summarizes much of his research in the journal *American Scientist,* and he has presented results at the annual meeting of the Society for Neuroscience in San Diego. His work departs sharply from the well-mined territory of humor analysis, in which scholars gather at conferences to discuss the ontology of Woody Allen or Monty Python and leave one with a distinct taste of sawdust in the mouth. Dr. Provine is not interested in formal comic material, or why some like Lenny Bruce and others Red Skelton, but in laughter as a universal social act.

"His work is extremely interesting, insightful," said Dr. William F. Fry, a psychiatrist at the Stanford University School of Medicine. "He's doing the sort of things that should have been done three hundred years ago." Dr. Fry is no joke himself, having studied the aerobic, physical and emotional benefits that accrue when a person laughs. One hundred laughs, he discovered, is equivalent to 10 minutes spent rowing.

Lest it appear that Dr. Provine is in the business of amusing himself and making strangers uncomfortable, he elaborates on the many quite serious questions that the study of laughter addresses. Laughter gives you a foothold on the neurobiology of behavior, he said. "It is species-typical, everybody does it and it is simple in structure, which gives you powerful leverage on the neurology behind it," he said.

"Looking at a common human behavior that is socially interesting gives us the opportunity to go back and forth between the neural circuitry and a higher social act," he said. Dr. Provine compares studying a simple system like laughter for clues to more complex types of human behavior to biologists' use of a simple organism like yeast or nematodes for delving into the thicket of genetics or brain development.

He points out that linguists and scientists who study speech are always searching for the deep underlying structure to language, those phonemes that might be recognized as language units by everybody, regardless of whether they are French, Chinese or New Guinean. But finding the common currency of language has proved quite difficult. "If you're interested in the mechanisms of speech, wouldn't it be useful to look at a vocalization that all individuals produce in the same way, such as laughter?" he asks rhetorically.

Laughter also has the useful property of being contagious, he said. When you hear laughter, you tend to start laughing yourself—hence the logic behind the sitcom's ubiquitous laugh track. And it is easy to assess whether the brain's circuitry for recognizing laughter has been activated, Dr. Provine said. "You don't need to use electrodes, or wait for clinical cases of brain lesions," to study laughter recognition, he said. All you have to do is see if the person laughs on hearing laughter.

The infectiousness of laughter also makes it a particularly interesting social activity to explore. Few behaviors, short of shouting "Fire" in a movie theater, can have such a dramatic, swelling impact on group behavior as can the burst of a merry chime of laughter. Indeed, Dr. Provine came to laughter research after studying another highly contagious human behavior: yawning.

Before he could hope to get at any neural circuitry, Dr. Provine first had to do the basics, starting with what a laugh looks like. He brought recorded samples of human laughter to the sound analysis laboratory at the National Zoo in Washington, where the usual subjects of research are bird songs and monkey screams. There he and colleagues generated laugh waveforms and laugh frequency spectrums. They determined that the average laugh consisted of short bursts of vowel-based notes—"haha" or "hehe"—each note lasting about 75 milliseconds and separated by rests of 210 milliseconds. Whether a person laughs with a shy giggle, a joyous musical peal or a braying hee-haw, "the key is the burst of vowel-like sounds produced in a regular rhythmic pattern," he said.

A typical laugh also has a decrescendo structure, starting strong and ending soft. A laugh played backward, going from low to high bursts, sounds slightly strange, almost frightening, and yet it is still clearly recognizable as a laugh, just as a bird song played backward would be; the same cannot be said for a human conversation played backward. "Laughter has more in common with animal calls than with what we think of as modern speech," Dr. Provine said.

Dr. Provine and his students also began gathering hundreds of episodes of everyday laughter. They were startled by the ordinariness of the comments that would elicit laughter. Equally surprising was how often people laughed at their own statements. The standard image of the comedian is the deadpan performer who hardly grins while the audience members convulse in laughter. But the average speaker chattering away laughs 46 percent more frequently than do those listening to the spiel.

There proved to be wide variations based on sex in the ratio of speaker-to-listener laughter. A man talking to a male listener laughs only slightly more than his companion will in response. If a woman is talking to a woman, she laughs considerably more than does her audience. By contrast, a male speaker with a female hanging on his words laughs 7 percent less often than does his appreciative hearer. And the biggest discrepancy of all is found when a woman speaks to a man, in which case she laughs 127 percent more than her male associate, who perhaps is otherwise occupied with planning a witty rejoinder.

Speakers and listeners alike abide by rules while laughing. Laughter almost never intrudes upon the phrase structure of speech. It never interrupts a thought. Instead, it occurs as a kind of punctuation, to reflect natural pauses in speech. This is true for listeners as well as the talkers: they do not laugh in the middle of a speaker's phrase, Dr. Provine said. And in fact to do so may be evidence of psychological abnormality; a crazy person may not wait for you to finish speaking before interrupting with a booming "HA!"

The lawfulness of the relationship between laughter and speech, said Dr. Provine, indicates a segregation of brain processes devoted to one or the other. "It suggests that you have mutually exclusive but interacting vocal processes," he said. "And it seems speech is dominant over laughter, because laughter does not intrude on speech."

What, then, is the purpose of all this lawfully punctuating chuckling? Laughter is, above all, a social act, Dr. Provine said. You are far more likely to talk to yourself while alone than laugh to yourself (unless you are watching television or reading, in which case you are engaged vicariously in a social event). Dr. Provine sees laughter as a within-group modulator, something designed to influence the tenor of an assemblage, to synchronize mood and possibly subsequent actions. He compares laughter to the barking of dogs, which coordinates disparate elements into a more or less like-minded team. Joyous laughter can help solidify friendships and pull people into the fold.

As with any group behavior, though, laughter has its menacing streak. The flip side of mirthful laughter may not be tears, but jeering, malicious laughter, used not to include people in one's group, but to exclude the laughable misfit. To make his point about the downside of laughter, and how it can turn deadly, Dr. Provine shows a clip from the movie *Goodfellas*, a scene in which the volatile Joe Pesci character laughs together with his fellow thugs before smashing a bottle of alcohol into a poor intruder's face.

Despots historically have feared the power of laughter; comedians during the Nazi era in Germany, for example, were kept on the Gestapo's shortest leash. "Fashions on laughter change, but one thing that stays the same is, you can't laugh at people in power," Dr. Provine said. The sanction holds for the personal as well as the political. Laugh at your boss, and you may be the recipient of that practical joke known as the little pink slip.

—NATALIE ANGIER, **February 1996**

A Universal Language

Laugh, and the world laughs with you—whatever the country, whatever the tribe. Anthropologists have never encountered a culture where people do not laugh to express merriment and sociability; even deaf people sometimes laugh out loud. Babies begin laughing at the age of two or three months, said Dr. William F. Fry of the Stanford University School of Medicine, just when the parents are getting fed up with all the fussing.

The rate of laughter picks up steadily for the next several years, until around the age of six or so, when the average child laughs 300 times a day.

After that, social training and the desire to blend in with one's peers conspire to damp liberal laughter. Estimates of how much adults laugh vary widely, from a high of 100 chuckles daily to a dour low of 15, but clearly adults lose their laughter edge along with the talent for finger-painting.

Some laughter authorities view that decline as a blow to the health of body and spirit. When you laugh robustly, you increase blood circulation, work your abdominal muscles, raise your heart rate and get the stale air out of your lungs; after a bout of laughter, your blood pressure drops to a lower, healthier level than before the buoyancy began.

Beyond such overt benefits, Dr. Lee S. Berk of the Loma Linda School of Public Health in California has also discovered subtler effects of laughter on the immune and neuroendocrine systems. He and his colleagues have learned that an hour spent laughing lowers levels of stress hormones like cortisol and epinephrine. At the same time, the immune system appears to grow stronger, the body's T cells, natural killer cells and antibodies all showing signs of heightened activity.

"It's no longer mystical," he said. "We've always heard that laughter is good for you, and now we're gathering the hard, serious stuff to show why this is so."

What remains to be demonstrated, though, is that any of these immune changes actually help fight or prevent disease.

Dr. John Morreall, a philosopher and laugh consultant who runs a firm called Humorworks out of Temple Terrace, Florida, tries to persuade businesses like Kodak and Xerox and, yes, also the Internal Revenue Service, of the financial value of laughter on the job. "It gets the creative juices flowing, it helps in problem solving, and it smooths the way during difficult times like annual evaluations or when giving criticism," he said. "The ability to think in funny ways and original ways can overlap." Punch lines and bottom lines likewise overlap: if you cannot afford to give your employees raises or job security, you can always give them the freedom to laugh.

Time Almost Buried Ancient Maya Secrets

WHEN THE AMERICAN LAWYER John Lloyd Stephens rediscovered the ancient Maya ruins in the early 19th century, the strange engravings on stone monuments fascinated and puzzled him. Of the city of Copan, he wrote, "One thing I believe, that its history is graven on its monuments. Who shall read them?"

Until 1960, it looked as if no one ever would. Archeologists had never found a Maya equivalent of the Rosetta stone, which was used to decode the ancient Egyptian inscriptions. Descendants of the Maya living in Central America had lost the knowledge of how to read the complex glyphs.

And scholars were getting nowhere because of an erroneous assumption: that the glyphs were all logographs, signs standing for entire words.

Then a Russian linguist, Yuri Knorosov, determined that a list of Maya signs preserved by early Spanish colonists did not represent letters of an alphabet, as had been assumed, but syllables. When the signs were joined in a block or elongated oval, they phonetically spelled words.

In a test, Mr. Knorosov found that two syllables formed a word almost identical to *kutz,* a modern Maya word for "turkey," and two other syllables created a glyph similar to *tzul,* a Maya word for "dog."

Western scholars were slow to accept the Russian's insight. By the time they did, in the 1960s, two other elements in the decipherment became apparent. Some glyphs, it was recognized, were not actual words but emblems either for places or some ruling families. In some cases, glyphs were indeed logographs, pictorial representations that stood not for a syllable but a whole word.

For example, the ancient Maya word *ahaw,* for "lord" or "ruler," often appeared as a logograph, a sketch resembling a distinguished man, or as a combination of curlicues and squiggles representing syllables that form the same word. Likewise, the word *pakal,* for "shield," can be shown as a depiction of a shield or by a combination of its syllabic elements. But the image of a bat turns out not to be the word for the animal, but a phonetic sign for a syllable.

David Stuart, a graduate student at Vanderbilt University and one of the foremost Maya epigraphers, said the Maya scribes seemed to enjoy wordplay, taking advantage of the language's flexibility and ambiguity. For example, the Maya word *kan* may mean "snake," "sky" or "four."

The 26-year-old Mr. Stuart is the wunderkind of epigraphy. The son of Maya archeologists, he was already deciphering the glyphs and reading stone monuments when he was only 12. Six years later, he became the youngest recipient ever of a MacArthur "genius" fellowship. He is working on his Ph.D. in Maya archeology at Vanderbilt.

Mr. Stuart is one of a handful of people who are struggling to decipher the Maya glyphs. They include Dr. Stephen Houston, a colleague at Vanderbilt, Dr. Linda Schele of the University of Texas, Dr. Peter Mathews of the University of Calgary, Dr. Floyd Lounsbury of Yale University and Dr. Nikolai Grube at Bonn University in Germany.

Without a Rosetta stone of any kind, their work has been slow and tedious. By now, about 1,000 of the Maya glyphs have been identified. "We don't know how much we still don't know," Mr. Stuart said.

—JOHN NOBLE WILFORD, November 1991

Language of Early Americans Is Deciphered

THERE ONCE WAS A WARRIOR by the name of Harvest Mountain Lord. He lived in a hot, humid land by the bend in a river that flowed into another river that ran to the sea. Many were the battles he fought and the blood rituals he endured, for this warrior was ruler of the people by the bend in the river.

The exploits of this warrior-king who lived more than 1,800 years ago—perhaps another culture's King Arthur or Napoleon—were recorded in hieroglyphics carved on a stone monument, or stela, found in 1986 in the Mexican state of Veracruz. On the face of the stela are a full-figure portrait of a richly attired man and a lengthy text telling of his rise to kingship through several years of warfare and elaborate accession rites presided over by a shaman.

From the moment of discovery, archeologists had felt sure that this could be one of the most important pre-Columbian monuments ever found. But they knew nothing of the story it told because they could not read the strange script. Now they can.

In a rare scholarly achievement, an archeologist and a linguist have deciphered the ancient writing system, called epi-Olmec. They determined that it was closely related to ancient Mayan writing, which has only recently been deciphered itself, and could be descended from the obscure hieroglyphics of the Olmecs, a pre-Mayan people along the Gulf of Mexico who developed one of the earliest major civilizations in the Americas.

Reporting their research in the journal *Science,* Dr. John S. Justeson of the State University of New York at Albany and Dr. Terrence Kaufman of

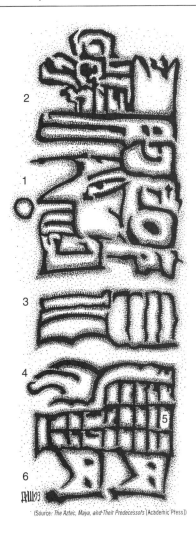

"Decoding a Name"
Researchers believe these glyphs refer to an ancient Olmec ruler. The head (1) wears a crown (2) that suggest the personage has been installed and consecrated as king. Beneath are images that give the king's name as "Lord Mountain Harvest." The knife (3) means "harvest." The next picture contains several images: vegetation (4), mountain (5) and lord (6). (Sources: hieroglyphics were rendered from an original photograph from The Center for Maya Research.)

(Source: *The Aztec, Maya, and Their Predecessors* [Academic Press])

the University of Pittsburgh said the decipherment "yields the earliest currently readable texts in Mesoamerica."

From evidence in the text, the basalt monument has been dated at A.D. 159. Since the earliest known complex writing systems in the hemisphere have been identified in the region archeologists call Mesoamerica, from central Mexico south into Central America, this would make the epi-Olmec text the earliest to be deciphered anywhere in America.

Dr. George E. Stuart, an editor of *National Geographic* magazine who is an archeologist and authority on Mayan culture, said the decipherment could part the curtain on an important but little-understood epoch in Mesoamerican cultural history. These are the centuries between the waning of the Olmec society, which flourished from 1200 to 500 B.C., and the literate Classic Period civilizations of the Mayans and others, beginning about A.D. 300.

The Olmecs were a particularly enigmatic people, known today mainly through their distinctive sculpture of huge human heads with masklike expressions and thick lips. Their heartland was in the Mexican states of Veracruz and Tabasco. Although they built no great cities, the Olmecs developed a wide trading network and left a legacy of art and religious ritual. The jaguar, a recurring theme in Olmec art, is a prominent figure in the stela text.

In a commentary accompanying the report, Dr. Stuart wrote, "The art of the Olmec and their Middle Preclassic neighbors appears to reflect many fundamental patterns seen in later Mesoamerican remains, including certain political and religious motifs and themes, the use of the calendar, and the beginning of writing—although examples of the latter are rare."

As appealing as the prospect of learning more about the Olmecs may be, Dr. Justeson, an archeologist and computer scientist, and Dr. Kaufman, an anthropological linguist, were cautious in their discussion of possible archeological implications of their work.

"Another Tale From Ancient Stone"
Researchers interpret this excerpt from the recently translated epi-Olmec stela as part of a description of a ritual. It instructs a king that "nine days later a jaguar spirit should be acquired."

4 dots plus a bar: the numeral 9.

Phonetic symbols that spell "day."

Jaguar spirit

"It should be acquired."

(Sources: Images rendered from an original photograph from the Center for Maya Research; interpretation by John Justeson)

"The script may itself descend from an Olmec hieroglyphic system, but too little of the Olmec script has been recorded to confirm or disprove a connection," they wrote.

The writing is called "epi-Olmec" because it was used by people living in the former Olmec lands, at least some of whom were probably Olmec descendants. The language represented in the text, the researchers found, is an early form of Zoquean, a branch of the Mixe-Zoquean language family still spoken by 100,000 to 140,000 rural people in the states of Veracruz, Tabasco, Chiapas and Oaxaca. The stela was found in the mud of the Acula River near the ranching village of La Mojarra, 25 miles inland from Alvarado on the Gulf Coast. Despite some breakage and erosion, the inscriptions were in remarkably good condition, and several teams of linguists got busy trying to make sense of the writing.

Describing their work in interviews, Dr. Justeson and Dr. Kaufman said several factors contributed to their success. For one thing, the length of the text enabled them to analyze the repetition of certain signs in different contexts and with different prefixes and suffixes, which helped establish the grammar. They also got clues to word meanings from calendar notations and comparisons with Mayan signs.

By contrast, the decipherment of ancient Egyptian in the early 19th century was a more straightforward task. The Rosetta stone bore a bilingual text, so semantic correspondences could be made between the unknown Egyptian and the known Greek, and the names of prominent historical figures like Cleopatra and Alexander were easily recognized.

Research by Dr. Kaufman as a graduate student 30 years earlier proved to be crucial. By studying the present Mixe-Zoquean languages of Mexico he reconstructed the grammar and basic vocabulary of their unknown ancestral language. "It was like reconstructing old Latin from our knowledge of modern Romance languages," Dr. Kaufman said.

The much shorter texts on a few other artifacts from the same region and time were also helpful. A statuette, for example, contained references to the same warrior-king and signs describing similar rituals. The statuette text ends with a shaman calling up the warrior's animal spirit and shows a person in the guise of an animal, wearing a duckbill mask and a cape of bird wings and claws.

By the completion of their research, the two scholars had identified in the 21 columns of stela hieroglyphs at least 150 signs of the epi-Olmec writing system. Most of them are abstract signs representing syllables, combinations of which make words. But more than 30 are logograms, graphic images representing the warrior-king, sunrise and stars, jaguars and a penis. As with the Mayans, the practice of bloodletting from the penis was a rite of renewal for the king and his fellow nobles.

Dr. Stuart characterized the stela as a sort of "political poster dealing with the accession to power of the individual portrayed." In this, he said, the text anticipates the content of later images and inscriptions from the Mayan area and elsewhere in Mesoamerica.

Few extinct written languages have been deciphered in this century, the most recent being Linear B, the Minoan script representing Mycenaean Greek, in the 1950s, and the Mayan hieroglyphics over the last three decades. Still defying scholarly analysis are Linear A from Crete and the early languages of the Indus River Valley and Easter Island.

"These things are so rare," Dr. Kaufman remarked, "I don't expect it to happen to me again."

Such decipherments can also be controversial. In his book *Breaking the Maya Code* (Thames and Hudson, 1992), Dr. Michael D. Coe, a Mayan scholar at Yale University, said the reaction by many archeologists to this "most exciting development in New World archeology in this century" has been rejection and even hostility. They have been exceedingly slow, he contended, in incorporating the deciphered texts in their interpretations of Mayan history.

Not many archeologists have examined the new research, but Dr. Justeson said, "We think that as they see the evidence they will find it over all to be convincing."

Dr. Richard Diehl, an archeologist at the University of Alabama at Tuscaloosa who plans excavations in near La Mojarra, said, "The crux of the problem is that there are still not enough texts to provide a critical data base for a thorough decipherment. We need to get more texts, and that's what I want to do."

With the greater interest in Mayan cultures to the south, the Olmec region has been little explored in recent decades. Dr. Justeson and Dr.

Kaufman said that to expand their decipherment and test their results they must wait for archeologists to dig up more epi-Olmec monuments bearing tales of other warrior-kings like Harvest Mountain Lord.

—JOHN NOBLE WILFORD, March 1993

In a Publishing Coup, Books in "Unwritten" Languages

JESUS SALINAS PEDRAZA, A RURAL schoolteacher in the Mexican state of Hidalgo, sat down to a word processor a few years back and produced a monumental book, a 250,000-word description of his own Indian culture written in the Nahnu language. Nothing seems to be left out: folktales and traditional religious beliefs, the practical uses of plants and minerals and the daily flow of life in field and village.

But it is more than the content that makes the book a remarkable publishing event, for Mr. Salinas is neither a professional anthropologist nor a literary stylist. He is, though, the first person to write a book in Nahnu (NYAW-hnyu), the native tongue of several hundred thousand Indians but a previously unwritten language.

Such a use of microcomputers and desktop publishing for languages with no literary tradition is now being encouraged by anthropologists for recording ethnographies from an insider's perspective. They see this as a means of preserving cultural diversity and a wealth of human knowledge. With even greater urgency, linguists are promoting the techniques as a way of saving some of the world's languages from imminent extinction.

Half of the world's 6,000 languages are considered by linguists to be endangered. These are the languages spoken by small societies that are dwindling with the encroachment of larger, more dynamic cultures. Young people feel economic pressure to learn only the language of the dominant culture, and as the older people die, the nonwritten language vanishes, unlike languages with a history of writing, like Latin.

Dr. H. Russell Bernard, the anthropologist at the University of Florida at Gainesville who taught Mr. Salinas to read and write his native language, said, "Languages have always come and gone. Neither the language of Jesus nor that of Caesar are spoken today. But languages seem to be disappearing faster than ever before."

In 30 years of field studies, Dr. Kenneth Hale, a linguist at the Massachusetts Institute of Technology and a leader in efforts to preserve endangered languages, said he had worked with at least eight that have become extinct and many others that are "seriously imperiled."

Dr. Michael E. Krauss, the director of the Alaska Native language Center at the University of Alaska in Fairbanks, estimates that 300 of the 900 indigenous languages in the Americas are moribund. That is, they are no longer being spoken by children, and so could disappear in a generation or two. Only two of the 20 native languages in Alaska are still being learned by children.

"Languages no longer being learned as mother tongue by children are beyond endangerment," Dr. Krauss said. Unless the current course is somehow reserved, he added, these languages "are already doomed to extinction, like species lacking reproductive capacity."

Dr. Krauss asks, "Should we mourn the loss of Eyak or Ubykh any less than the loss of the panda or California condor?"

At a symposium a few years ago in Mexico on "The Politics of Linguistic Revitalization," representatives of Latin American Indians passed a resolution saying, "The loss of a language in the world means the disappearance of the cultural heritage transmitted by it, and the truncation of an alternate route of cultural development for humankind."

In an effort to preserve language diversity in Mexico, Dr. Bernard and Mr. Salinas decided in 1987 on a plan to teach the Indian people to read and write their own languages using microcomputers. They established a native literacy center in Oaxaca, Mexico, where others could follow in the footsteps of Mr. Salinas and write books in other Indian languages.

The Oaxaca center goes beyond most bilingual education programs, which concentrate on teaching people to speak and read their native languages. Instead, it operates on the premise that, as Dr. Bernard decided, what most native languages lack is native authors who write books in their own languages.

"Without popular literacy, all but a few endangered languages will soon disappear," Dr. Bernard said. "And when nonwritten languages disappear, they disappear forever."

Mr. Salinas set the example. He had grown up speaking both Spanish and Nahnu, one of the more widely spoken of the 56 Indian languages in Mexico. The Nahnu Indians are also known as the Otomi, a name they now reject because of its pejorative connotations in Mexican Spanish.

While conducting research in Mexico in the 1970s, Dr. Bernard taught Mr. Salinas to read and write the Nahnu language, using a modified version of an alphabet that had been developed by missionaries. The Summer Institute of Linguistics, a worldwide evangelical Protestant organization based in Dallas, has for decades developed the alphabets for hundreds of Indian languages.

Virtually all indigenous languages now have rudimentary writing systems, including alphabets, a dictionary and some grammar. In Nahnu and many other languages, however, the only book is a Bible translation.

Next Dr. Bernard adapted computer software for writing Nahnu on a word processor and for printing manuscripts and books, which would not have been affordable through traditional publishing. In this way, Mr. Salinas first produced a book of Nahnu folktales and, two years ago, his magnum opus on Nahnu culture. A translation in English, by Dr. Bernard, is entitled, *Native Ethnography: A Mexican Indian Describes His Culture*.

He wrote of the trapping and cooking of squirrels, which many people still believe were first human beings who sinned and so are symbols of bad luck; the uses of the mineral lime in medicine (for burning off warts) and in making tortillas and cooking vegetables; the arrival of foreigners, as remembered in oral histories; rituals of passage; music; and local fiestas.

There is a long tradition of training people of a remote culture to be what anthropologists call "informants." These people would describe the legends and customs and sometimes dictate their autobiographies. But these were usually taped interviews with an anthropologist and not the direct work of an indigenous person.

At the Oaxaca center, run by Mr. Salinas, Indians spend up to three months learning the technology for producing works in their previously nonwritten languages. They are mostly bilingual schoolteachers, such as Mixtecs learning to use a Mixtec-Spanish word processor. Josefa González, the wife of Mr. Salinas, is a Mixtec Indian who is writing her own book on Mixtec culture.

The center is supported by the Jessie Ball du Pont Foundation and sponsored by Mexico's Ministry of Education, the Inter-American Indian Institute of the Organization of American States, the Center for Advanced Studies in Social Anthropology in Oaxaca and the University of Florida at Gainesville.

So far, the Oaxaca authors have produced works in Chatino, Amuzgo, Chinantec and Mazatec, as well as Nahnu and Mixtec. Other books in Tzotzil and Zapotec are in the works. In every case, the trainees wrote directly in their own languages, using word processors with specially developed computer programs. Such microcomputer technology is available at relatively low cost, Dr. Bernard said, "so that virtually every nonliterate people of the world can develop their own machine-readable data bank about their own culture, and can have their own independent publishing capability in support of native ethnography."

The Oaxaca project's influence is spreading. Impressed by the work of Mr. Salinas and others, Dr. Norman Whitten, an anthropologist at the University of Illinois, arranged for schoolteachers from Ecuador to visit Oaxaca and learn the techniques.

Now Ecuadoran Indians have begun writing about their cultures in the Quichua and Shwara languages. Others from Bolivia and Peru are learning to use the computers to write their languages, including Quechua, the tongue of the ancient Incas, still spoken by about 12 million Andean Indians.

Anthropologists in Cameroon have introduced the technology in that African country of more than 200 tribal languages. At the end of a two-week training course, Dr. Bernard said, each of the five speakers of the Kom language would turn on the computer in the morning and start typing Kom without hesitation. Anything they could say, they could write.

Dr. Bernard emphasized that these native literacy programs are not intended to discourage people from learning the dominant language of their country as well. "I see nothing useful or charming about remaining monolingual in any Indian language if that results in being shut out of the national economy," he said.

Anthropologists acknowledge that bilingualism can be a sensitive political issue. In places struggling to create a workable nation, the preservation of separate languages can be seen as potentially divisive, setting one group against another.

But anthropologists and leaders of native cultures generally favor bilingualism because it preserves a distinctive culture, which often gives the people more power than if they were completely assimilated. A distinctive language can be a strong force in establishing cultural uniqueness, or ethnicity, which in turn can reinforce a group's claim to a share of power.

Such efforts may be too late for hundreds of languages and dwindling cultures, anthropologists and linguists say, but in time to help preserve some of the cultural heterogeneity of the planet.

—JOHN NOBLE WILFORD, December 1993

Indians Striving to Save Their Languages

AT THE AGE OF 88 and blind in one eye, James Jackson, Jr., keeps a crystal clear memory of a tiny "linguistic" skirmish in a continental campaign that has brought most of North America's Indian languages to the brink of extinction.

"The teacher at the Indian school grabbed my friend by the arm and said, 'You're speaking your language—I'm going to wash your mouth out with soap,'" Mr. Jackson recalled. "That's where we lost it."

Eight decades later, Mr. Jackson told his story, in English, to a small circle of Hupa language students at the Hoopa Community Center. Although the tribe has about 2,000 members, the room contained the four people who make up about half of the world's fluent Hupa speakers: Mr. Jackson, his younger sister, Minnie, and two elderly friends. Two others died in February.

Despite five centuries of population decline, assimilation and linguistic oppression, most of North America's Indian languages have survived to the end of the 20th century. Of the approximately 300 Indian languages that existed when Europeans first arrived in what is now the United States and Canada, 211 are still spoken today.

But with the impact of television and radio and increased mobility among Indians, North America's native languages are suffering their sharpest free fall in recorded history.

Of the 175 Indian languages still spoken in the United States, only 20 are still spoken by mothers to babies, said Michael E. Krauss, a linguist at the University of Alaska who surveys native languages. In contrast, 70 languages are spoken only by grandparents, and 55 more are spoken by 10 tribal members or fewer.

"This is a major American tragedy that people are generally in a state of denial about," Dr. Krauss said.

Noting that the federal government spends only $2 million a year to save endangered Indian languages, he said that under the Endangered Species Act, "we are spending one million dollars a year per Florida panther to save the species."

Belonging to 62 language families, American Indian languages are as dramatically different as German, Chinese and Turkish.

With the rise of a global economy and increased communications, about half of the world's 6,000 languages are expected to disappear over the next century. Among American Indians, that process is unfolding today.

"We just gave a grant to study Klamath," said Douglas H. Whalen, a Yale University linguist who directs a new nonprofit group, the Endangered Language Fund. "When the proposal was made, Klamath had two speakers. Now it is down to one."

Rapid erosion is also affecting the largest tribes.

In Arizona among the Navajo, the most populous tribe in the United States, the portion of native speakers among first graders has dropped to 30 percent today from 90 percent in 1968.

In Montana, the 9,300 enrolled Crow members display a classic Indian linguistic profile: 77 percent of Crow Indians over 66 years of age speak the language, while only 13 percent of preschoolers do.

On paper, California has the most linguistic diversity in the nation: 50 Indian languages are still spoken there, down from 80 before Europeans arrived.

"But not a single one of those languages is now being spoken natively by children," said Leanne Hinton, a linguistics professor at the University of California at Berkeley. "We are heading toward a state where we will have no native speakers of any of the California languages in ten or twenty years. We are entering an age when speakers of the California languages will be learning in school, or as adults, rather than at home."

As the language circle at the Hoopa Community Center suggests, however, there is a belated movement among American Indians to rescue their languages from extinction.

"It's part of our culture," said Daniel Ammon, a Hupa high school teacher who is one of several dozen adults studying the tribe's complex language. "It contains how a Hupa person views the world. To lose the language would be to lose our identity. I will talk to my kids all the time in Hupa."

At the regional high school in Hoopa, a town of 1,000 people on a bend of the Trinity River in northwestern California, classes started last fall in the three languages of Indians in the area: Hupa, Karuk and Yurok.

"I want to know Karuk because it is my language, because I want to teach it to my children," Nisha Supahan, 15, said after class as her twin sister, Elaina, giggled in assent.

Their 27-year-old Karuk teacher, Susan Smith, contrasted their attitude with her detribalized upbringing.

"I never heard my language as a child," Ms. Smith said. "I didn't even know how to pronounce my tribal name."

Mr. Ammon and Ms. Smith learned their native language through an innovative effort to stave off linguistic extinction. Since 1992, the Native California Network, a nonprofit group in Visalia, in the state's south-central region, has sponsored 50 "apprentices" to undergo intensive language immersions, sometimes for up to 500 hours, with "masters," tribal elders who speak the language.

The language revival effort is taking many forms. Last year the Crow Tribal Council adopted resolutions declaring Crow the official language of the reservation, honoring fluent speakers as "tribal treasures" and encouraging all tribal members to speak the language.

Elsewhere in Montana, the Northern Cheyenne are offering tribal children a summer language camp, taught by the five elders who still speak Cheyenne fluently. In Missoula, summer language classes are offered in Blackfeet. Across Montana, a recent state decision to ease the certification of Indian language instructors has led to a surge in language instruction.

Idaho State University now offers Shoshoni for foreign language credit.

"At least one quarter of the thirty tribal colleges now require language study," David Cournoyer, a director of the American Indian College Fund, said in Denver. "Today, twenty-five different languages are taught, plus Plains Indian sign language."

In Connecticut, the Mohegan and Pequot are studying written records in their languages in an effort to revive languages that have not been spoken since the early 1900s. Thanks to the work of missionaries and anthropologists, virtually all of the Indian languages in North America have dictionaries and written texts.

While official language extermination policies have stopped, the main threat today, said Dr. Krauss, the University of Alaska linguist, is "the cultural nerve gas of television."

STATUS REPORT: Still Spoken Here, but for How Long?

Although 175 Indian languages are still spoken in the United States, and all of them are in danger of disappearing, some are more endangered than others.

Some children are being raised in the language; still spoken by parents and the elderly: **20 languages**
  Examples: The majority of these languages are in New Mexico and Arizona, such as NAVAJO, WESTERN APACHE, HOPI, ZUNI and HAVASUPAI-HUALAPAI. Others include CHOCTAW in Mississippi, YUPIK in Alaska, CHEROKEE in Oklahoma and LAKOTA-DAKOTA in the northern Plains.

Spoken by parents and the elderly: **30 languages**
  Examples: CROW and CHEYENNE in Montana, MESQUAKIE in Iowa, JICARILLA APACHE in New Mexico and GWICH'IN in northeastern Alaska.

Spoken almost entirely by the elderly: **70 languages**
  Examples: TLINGIT in the Alaskan panhandle, PASSAMAQUODDY in Maine, WINNEBAGO in Nebraska, ONEIDA and SENECA in upstate New York, HIDATSA in North Dakota, COMANCHE in Oklahoma, YUMA in California, NEZ PERCE in Oregon and KALISPEL, YAKIMA and MAKAH in Washington.

Spoken by fewer than 10 elderly tribal members: **55 languages**
  Examples: EYAK in south-central Alaska, PENOBSCOT in Maine, TUSCARORA in New York, MANDAN in North Dakota, Delaware, Iowa, PAWNEE and WICHITA in Oklahoma, CHEHALIS, CLALLAM, COWLITZ and SNOHOMISH in Washington, OMAHA in Nebraska and WASHOE in California.

(Source: Dr. Michael Krauss, Director of the Alaska Native Language Center, University of Alaska Fairbanks)

Putting electronic communications to work, the Hopi of Arizona have expanded Hopi language radio broadcasting, the Choctaw of Oklahoma have produced native language video dramas, the Sioux of South Dakota maintain a Lakota language Internet chat room, and the Skomish of Washington have produced a Twana language CD-ROM.

"There has been an almost total inversion in attitudes toward the native language," said Victor Golla, a linguistics professor at nearby Humboldt State University, who started visiting here 30 years ago. "Before, people were unconcerned about their native language. Now there is a very strong feeling among almost all the people that the loss of their language would be a tragic and very damaging thing."

Indians interested in reviving traditional ways say they cannot pray to their ancestors in English.

"A number of people have learned how to pray in their language," said Ms. Hinton, the Berkeley linguistics professor, who runs a summer program for Indians in California seeking to revive their languages from recorded field notes and tapes. "They are starting to reinvent their languages so they can pray at ceremonies and funerals."

Linguists caution that the language revival movement may only delay inevitable extinctions. But here in Hoopa, a change can already be felt.

"Before, on the bus, I used to say to my sister in Karuk, 'Look at that guy's shirt,' and nobody knew what we were talking about," Nisha Supahan said. "Now that's not true anymore."

—JAMES BROOKE, April 1998

6

THE LATEST FROM THE FIELD

Perhaps the most dramatic recent event in the field of language and linguistics has been the discovery of a high level gene for language. Biologists hope that study of the gene will give them access to the many other genes that doubtless underlie the human communication system.

The gene came to attention because it is defective in about half the members of a large London family. They lack the ability for fluent speech but otherwise appear mostly normal, suggesting that the variant gene is specific for language.

Supporting the idea that the gene underlies a specific human ability, another group of researchers found that the human version of the gene not only differs from the version found in chimpanzees, but is universal in the human population, so far as is known. This is just what would be expected for a gene that conferred such a decided advantage as the gift for language. Two of the following articles describe this development.

Another discusses the publication of Joseph Greenberg's last great work of synthesis, his argument that a great swath of languages spoken from Spain to Japan all belong to a single family, which he calls Eurasiatic. Most historical linguists refuse to make such overarching syntheses, believing languages change so rapidly that similarities are completely washed out after a few thousand years. Greenberg's syntheses, however, have received corroboration from an independent source, the history of human populations as mapped out by geneticists. The population splits as reconstructed from DNA turn out to match quite closely the language splits inferred by Greenberg.

Dr. Greenberg died in 2001, his work being widely honored by population geneticists, but still disdained by many in his own discipline.

A fourth article discusses a new insight into the rules for language made by Mark Baker, a disciple of the linguist Noam Chomsky. Baker believes he has recognized the basic elements that form the deep structure of all languages.

As linguists work down from the syntax of languages, and geneticists work up from the information content of the genome, there is the hope that the two will meet and illuminate each other's problems. Fluently spoken, syntactical language is the only ability that clearly sets humans apart from other animals. But for language, man would be just another species of African ape. The understanding of language is inseparable from the understanding of human uniqueness.

———————

What We All Spoke When the World Was Young

IN THE BEGINNING, there was one people, perhaps no more than 2,000 strong, who had acquired an amazing gift, the faculty for complex language. Favored by the blessings of speech, their numbers grew, and from their cradle in the northeast of Africa, they spread far and wide throughout the continent.

One small band, expert in the making of boats, sailed to Asia, where some of their descendants turned westward, ousting the Neanderthal people of Europe and others east toward Siberia and the Americas.

These epic explorations began some 50,000 years ago and by the time the whole world was occupied, the one people had become many. Differing in creed, culture and even appearance, because their hair and skin had adapted to the world's many climates in which they now lived, they no longer recognized one another as the children of one family. Speaking 5,000 languages, they had long forgotten the ancient mother tongue that had both united and yet dispersed this little band of cousins to the four corners of the earth.

So might read one possible account of human origins as implied by the new evidence from population genetics and archaeology. But the implication that all languages are branches of a single tree is a subject on which linguists appear strangely tongue-tied.

Many deride attempts to reconstruct the family tree of languages beyond the most obvious groupings like the Romance languages and Indo-European. Their argument is that language changes too fast for its roots to be traced back further than a few thousand years. If any single language ever existed, most linguists say, it is irretrievably lost.

But one scholar in particular, Dr. Joseph H. Greenberg of Stanford University, has defied this ardent pessimism. In the course of a long career, he has classified most of the world's languages into just a handful of major groups.

Though it remains unclear how these superfamilies may be related to one another, he has identified words and concepts that seem common to them all and could be echoes of a mother tongue.

And this month, at the age of 84, Dr. Greenberg is publishing the first of two volumes on Eurasiatic, his proposed superfamily that includes a swath of languages spoken from Portugal to Japan.

Like the biologist E. O. Wilson, Dr. Greenberg is that rare breed of academic, a synthesizer who derives patterns from the work of many specialists, an exercise the specialists do not always welcome. But though biologists came to acknowledge the pioneering value of Dr. Wilson's work, linguists have reached no such consensus on that of Dr. Greenberg.

Will he one day be recognized as having done for language what Linnaeus did for biology, as his Stanford colleague and associate Dr. Merritt Ruhlen believes, or is his work more fit, as one critic has urged, to be "shouted down"?

Dr. Greenberg is by no means an outcast from his profession. He is one of the very few linguists who are members of the National Academy of Sciences, the country's most exclusive scientific club. His work on language typology (universal patterns of word order) is highly regarded. Somewhat puzzlingly, his fellow linguists generally accept his work on the relationships among African languages but furiously dispute his ordering of American Indian languages, even though both classifications were achieved with the same method.

Dr. Greenberg's work is of considerable interest to population geneticists trying to reconstruct the path of early human migrations by means of genetic patterning in different peoples. Although genes and languages are not bequeathed in the same way, both proceed in a series of population splits.

"We have found a lot of significant correspondences between what he says and what we see genetically," said Dr. Luca Cavalli-Sforza, a leading population geneticist at Stanford. In his view, the majority of linguists are not interested in the evolution of language. They "have attacked Greenberg cruelly, and I think frankly there is some jealousy behind it because he has been so successful," Dr. Cavalli-Sforza said.

In a windowless office lined with grammars and dictionaries of languages from all over the world, Joseph Greenberg fishes in the plastic shopping bag that is serving as his briefcase. He pulls out one of the handwritten notebooks that are the key to his method of discovering language relationships. Down the left hand margin is a list of the languages being compared. Along the top are names of the vocabulary words likely to yield similarities.

His method, which he calls mass or multilateral comparison, is to compare many languages simultaneously on the basis of 300 core words in the hope that they will sort themselves into clusters representative of their historical development. Many linguists believe such an exercise is futile because words change too quickly to preserve any ancestry older than 5,000 years or so.

"They sell their own subject short," Dr. Greenberg says. "Certain items in language are extremely stable, like personal pronouns or parts of the human body."

Born in Brooklyn in 1915, he was interested in language almost from birth. His father spoke Yiddish and his mother's family German. "I was brought up to believe Yiddish was an inferior language because my father's relatives got invited to the house as seldom as possible," he said. Hebrew school exposed him to a fourth language. He had a good enough ear that an alternative career as a professional pianist beckoned.

But anthropology won out. After doctoral studies at Northwestern, he did fieldwork on the pagan cults of the Hausa-speaking people of northern Nigeria before deciding that his true interest lay in linguistics.

At the time, there was no agreement on the history of African languages. "So I started in a simple-minded way," Dr. Greenberg said. "I took common words in a number of languages and saw if the languages fell into groups." He found that he could reduce all the continent's languages first to 14 and later to 4 major clusters.

In a 1955 article, he described these as Afro-Asiatic, which includes the Semitic languages of Arabic and Hebrew, as well as ancient Egyptian, and is spread across Northern Africa; Nilo-Saharan, a group of languages spoken in Central Africa and the Sudan; Khoisan, which includes the click languages of the south; and Niger-Kordofanian, a superfamily that includes everything in between, including the pervasive Bantu languages.

After a decade of controversy, Dr. Greenberg's African classification became widely accepted. "But then a lot of people said I had gotten the correct results with the wrong method," he said.

Method is the formal issue that divides Dr. Greenberg from his critics. They say that the only way to prove that a group of languages is related is by establishing regular rules governing how words change as one language morphs into another. The 'p' sounds in ancestral Indo-European, for example, change predictably into 'f' in German and English. Mere similarities between the words in different languages, like those on which Dr. Greenberg relies, fall far short of proof, his critics say, because the similarities could arise from chance or borrowing.

Because of the looseness of sound and meaning that Dr. Greenberg allows in claiming similarities, his data "do not rise above the level of chance," said Dr. Sarah Thomason, a linguist at the University of Michigan.

Dr. Brian D. Joseph of Ohio State University, who studies Nostratic, a proposed language superfamily similar to Euroasiatic, described Dr. Greenberg as "a romantic" for believing his methods could retrieve long lost languages.

Dr. Lyle Campbell, of the University of Canterbury in New Zealand and the author of a textbook on historical linguistics, said that rigorous proof was necessary because languages changed so fast, and that Dr. Greenberg's methods were "woefully inadequate."

To Dr. Greenberg and his colleague Dr. Ruhlen, the critics' requirement for establishing regular rules of sound change defies both common sense and history. The sound regularities in Indo-European, they say, were not detected until after the languages had been grouped by inductive methods similar to Dr. Greenberg's. The insistence on demonstrating sound-change regularities, in their view, has thwarted any further reconstruction of language families.

"It's a misguided perfectionism that is so perfect they have had no result," Dr. Ruhlen said. His and Dr. Greenberg's aim is to establish the probable links from which the full history of human language can be inferred.

"The ultimate goal," Dr. Greenberg said in concluding his 1987 book *Language in the Americas* (Stanford University Press), "is a comprehensive classification of what is very likely a single language family. The implications of such a classification for the origin and history of our species would, of course, be very great."

Because the Americas have been inhabited only recently, at least as compared with Africa, it would be surprising to find a larger number of language groups, and Dr. Greenberg decided there were only three, even though other linguists posit 100 or so independent stocks.

Amerind is the vast superfamily to which, in his view, most native languages of North and South America belong. The other two clusters are Na-Dene, a group of languages spoken mostly in Alaska and northeast Canada, and Eskimo-Aleut, spoken across northern Alaska and Canada.

One striking feature that unites the Amerindian languages of both Americas, in Dr. Greenberg's view, is the use of words starting in 'n' to mean I/mine/we/ours and words beginning in 'm' to mean thou/thine/you/ yours. Not every language shows this pattern, but almost every Amerindian language family has one or more languages that have it, suggesting that all are derived from an original language in which first and second person pronouns started this way.

In the course of classifying the languages of the Americas, Dr. Greenberg realized that their major families were related to languages on the Eurasian continent, as would be expected if the Americas had been inhabited by people migrating through Siberia. Na-Dene, for example, is related to an isolated Siberian language known as Ket.

To help with the American classification, Dr. Greenberg started making lists of words in languages of the Eurasian land mass, particularly personal pronouns and interrogative pronouns.

"I began to see when I lined these up that there is a whole group of languages through northern Asia. I must have noticed this 20 years ago. But I realized what scorn the idea would provoke and put off detailed study of it until I had finished the American languages book," he said.

Thirteen years later, Dr. Greenberg has now classified most of the languages of Europe and Asia into the superfamily he calls Eurasiatic. Its seven living components are Indo-European (examples are English, Russian, Greek, Iranian, Hindu); Uralic (Hungarian, Finnish); Altaic (Turkish, Mongolian); the Korean-Japanese-Ainu group; Eskimo-Aleut; and two Siberian families known as Gilyak and Chukotian.

His concept of Eurasiatic was derived independently but overlaps with the proposed Nostratic superfamily, the theory of which has been developed in the last 30 years by Russian linguists.

At first sight it may seem hard to believe that languages as different as English and Japanese, say, share any commonalities. But in his new book on the grammar of Eurasiatic (a second volume on vocabulary is in progress), Dr. Greenberg has found many elements that he argues knit the major Eurasian language families into a single group.

Words beginning in 'm,' for example, are found in every Eurasiatic family to designate the first person (English: me; Finnish: mina; proto-Altaic: min; Old Japanese: mi). Every branch of Eurasiatic, Dr. Greenberg says, uses n-words to designate a negative, from the no/not of English to the -nai ending that makes Japanese verbs negative.

Every branch uses 'k' sounds to indicate a question. In Indo-European, many Latin interrogatives begin qu-, as in quid pro quo. In Finnish, -ko is added to a verb to indicate a question. In Japanese the same role is played by -ka. The word for 'who?' is kim in Turkish, kin in Aleut.

If Dr. Greenberg's Eurasiatic proposal is at first no more favorably received than his Amerindian classification, he will not be surprised. "A fair part of my publications is just polemics," he says, with an air of resignation.

Meanwhile, Dr. Ruhlen believes that if the Eurasiatic grouping is accepted, the world's 5,000 languages can be seen to fall into just 12 superfamilies.

How these in turn might be related to a single mother tongue remains to be seen. But several years ago, Dr. Greenberg identified a possible global etymology derived from the universal human habit of holding up a single finger to denote one.

In the Nilo-Saharan languages the word tok, tek or dek means one. The stem tik means finger in Amerind, one in Sino-Tibetan, 'index finger' in Eskimo and 'middle finger' in Aleut. And an Indo-European stem deik, meaning to point, is the origin of daktulos, digitus, and doigt—Greek, Latin and French for finger—as well as the English word digital.

No one has pointed more clearly at the one language than Joseph Greenberg.

Echoes of a Mother Tongue?

Languages evolve so quickly that living versions are unlikely to bear obvious traces of the mother tongue from which they all may derive. But some words, like those for parts of the body, are more stable than others. Almost all of Dr. Greenberg's major language families contain languages in which a 'tik' or 'dik'-like word is used for the concept of one or finger or point. In English, the word is digit or digital.

LANGUAGE FAMILY	FORM	MEANING
Niger-Kordofanian	"Dike"	One
Nilo-Saharan	"Tekdektok"	One
Khoisan	[Absent]	
Afro-Asiatic	"Tak"	One
Dravidian	[None]	
Indo-European	"Deik"	To point
Uralic-Yukaghir	"Otik"	One
Altaic	"Tek"	Only
Chukchi-Kamchatkan	[Unclear]	
Eskimo-Aleut	"Tik"	Index, middle finger
Sino-Tibetan	"Tik"	One
Austric	"Tikting"	Hand, arm
Indo-Pacific	"Dik"	One
Australian	[Absent]	
Amerind	"Tik"	Finger, one
Na-Dene	"Tikhi"	One

(Source: *On the Origin of Languages* by Merritt Ruhlen)

—NICHOLAS WADE

Expert Says He Discerns "Hard-Wired" Grammar Rules

IN 1981 THE LINGUIST NOAM CHOMSKY, who had already proposed that language was not learned but innate, made an even bolder claim.

The grammars of all languages, he said, can be described by a set of universal rules or principles, and the differences among those grammars are due to a finite set of options that are also innate.

If grammar were bread, then flour and liquid would be the universal rules; the options—parameters, Dr. Chomsky called them—would be things like yeast, eggs, sugar and jalapeños, any of which yield a substantially different product when added to the universals. The theory would explain why grammars vary only within a narrow range, despite the tremendous number and diversity of languages.

While most linguists would now agree that language is innate, Dr. Chomsky's ideas about principles and parameters have remained bitterly controversial. Even his supporters could not claim to have tested his theory with the really tough cases, the languages considered most different from those the linguists typically know well.

But in a new book, Dr. Mark C. Baker, a linguist at Rutgers University whose dissertation was supervised by Dr. Chomsky, says he has discerned the parameters for a remarkably diverse set of languages, especially American-Indian and African tongues.

In the book, *The Atoms of Language: The Mind's Hidden Rules of Grammar* (Basic Books, 2001), Dr. Baker sets forth a hierarchy of parameters that sorts them according to their power to affect and potentially nullify one another.

Just as the periodic table of elements illustrates the discrete units of the physical world, Dr. Baker's hierarchy charts the finite set of discrete factors that create differences in grammars.

That these parameters can be organized in a logical and systematic way, Dr. Baker says, suggests that there may be some deeper theory underlying them, and that the hierarchy may even guide language acquisition in children.

The hierarchy is not the same as a family tree, which illustrates the historical relations among languages—for example, Italian, French, Spanish and their mother tongue, Latin. Nor does it have anything to do with the way words vary from language to language. Instead, Dr. Baker analyzes grammar—the set of principles that describe the order in which words and phrases are strung together, tenses added and questions formed. Dr. Baker, like Dr. Chomsky, believes these instructions are hard-wired into humans' brains.

His most spectacular discovery is that the grammars of English and Mohawk, which appear radically different, are distinguished by just a single powerful parameter whose position at the top of the hierarchy creates an enormous effect.

Mohawk is a polysynthetic language: its verbs may be long and complicated, made up of many different parts. It can express in one word what English must express in many words. For example, "Washakotya'tawitsherahetkvhta'se'" means, "He made the thing that one puts on one's body ugly for her"—meaning, he uglified her dress.

In that statement, "hetkv" is the root of the verb "to be ugly." Many of the other bits are prefixes that specify the pronouns of the subject and object. Every verb includes "each of the main participants in the event described by the verb," Dr. Baker writes. In all, Mohawk has 58 prefixes, one for each possible combination of subject, object and indirect object.

Dr. Baker says the polysynthesis parameter is the most fundamental difference that languages can have, and it cleaves off Mohawk and a few other languages—for example, Mayali, spoken in northern Australia—from all others. That two such far-flung languages operate in the same way is more evidence for the idea that languages do not simply evolve in a gradual or unconstrained fashion, Dr. Baker says.

Expert Says He Discerns 'Hard-Wired' Grammar Rules

At the next junction in the hierarchy, two parameters are at work: "optional polysynthesis" (in which polysynthetic prefixes are possible, but not required) and "head directionality," which dictates whether modifiers and other new words are added before or after existing phrases. In English, new words are at the front. For example, to make a prepositional phrase "with her sister," the preposition goes before the noun. In Lakota, a Sioux language, the reverse is true. The English sentence "I will put the book on the table" reads like this in Lakota: "I table the on book the put will." Japanese, Turkish and Greenlandic are other languages that opt for new words at the end of phrases, while Khmer and Welsh have the same setting as English.

In all, Dr. Baker and others have identified about 14 parameters, and he believes that there may be 16 more.

Dr. Baker's work is by no means universally accepted. Dr. Robert Van Valin, a professor of linguistics at the State University of New York at Buffalo, says the findings rest on a questionable assumption: that there is a universal grammar.

"What they're doing in that whole program is taking English-like structures and putting the words or parts of words of other languages in those structures and then discovering that they're just like English," he said.

Dr. Karin E. Michelson, an associate professor of linguistics at SUNY Buffalo, who also disagrees with the Chomskyan approach, said after reviewing Dr. Baker's Mohawk work that some of the sentences he selected seemed artificial.

Dr. Baker acknowledged that some of the longer words in his study were "carefully engineered," but he said the parameter still held up using more common examples of Mohawk. He said using only examples from real discourse restricted the kind of analysis that linguists could do.

"It would be like constraining a physicist to learn about gravity without ever building a vacuum tube," Dr. Baker said.

Other linguists, however, say they are excited by Dr. Baker's work. "He's a very influential linguist, and my guess is that this will provide insights and will spawn research for the next few years," said Dr. Stephen Crain, a professor of linguistics at the University of Maryland.

If Dr. Baker's theory is correct, a further question is how the parameters of grammar are set as a child learns language. Does a child in an English-speaking environment start at the top of the hierarchy, somehow grasp that polysynthesis is not at work, and then move on to the next level in the hierarchy?

Dr. Baker also wonders why, if the brain is hard-wired for grammar, it leaves the parameter settings unspecified. Why aren't they hard-wired, too?

Humans are assumed to have language in the first place because it allows them to communicate useful information to others. But perhaps, Dr. Baker speculates, language is also a tool of cryptography—a way of concealing information from competitors.

In that case, he went on, "the parameters would be the scrambling procedures."

—Brenda Fowler

Researchers Say Gene Is Linked to Language

A TEAM OF GENETICISTS and linguists say they have found a gene that underlies speech and language, the first to be linked to this uniquely human faculty.

The discovery buttresses the idea that language is acquired and generated by specific neural circuitry in the brain, rather than by general brain faculties.

The gene, which joins a handful known to affect human behavior, is of particular interest because its role is to switch on a cascade of other genes in the developing brain of the fetus. Biologists hope that by identifying these "downstream" genes, they may be able to unravel the genetic basis of human language.

The discovery may also help scientists answer the vital question of when language evolved and whether the power it gave modern humans was the primary reason they flourished and spread rapidly around the world.

Some scientists, however, say they believe the gene may be less specific to language than it seems. So the new finding could simply fuel a longstanding debate among linguists as to whether the brain handles language through mechanisms specifically dedicated to the task or through a more general system.

The new discovery is described in today's issue of *Nature* by Dr. Anthony P. Monaco of the University of Oxford in England and colleagues.

The gene first came to light through study of a large family, half of whose members have trouble pronouncing words properly, speaking grammatically and making certain fine movements of the lips and the tongue. Asked to speak a repetitive sound like "pataca pataca pataca," they will stumble over each iteration. Outsiders have trouble understanding them when they speak, and family members have difficulty understanding one another. Some of the affected members, though not all, seem normal otherwise, suggesting that a specific impairment of speech and language is the root of their problem.

The new study shows that all the affected members have inherited a mutation, or variant piece of DNA, in a specific gene. The mutation affects a single unit in the 6,500 units of DNA that make up the gene. So delicate is the human genetic programming that this minuscule change suffices to sabotage the whole faculty of speech and language.

The carriers of this variant gene resemble other people who have impairments of language. They came to researchers' attention in 1990 only because there were so many of them, all related and all living in the same area of London. The family now has 29 members in three generations, 14 of whom have the disorder.

The first linguist to study the family, Dr. Myrna Gopnik of McGill University in Montreal, reported in 1990 that affected members were unable to change the tense of verbs correctly, a finding that provoked considerable stir in the linguistic world because it implied the existence of genes for grammar.

But in a later study of the family, Dr. Faraneh Vargha-Khadem of the London Institute of Child Health identified a much wider range of speech and language deficits, and some effects on general intelligence. The variant gene "affects speech, but with knock-on effects in nonverbal ability," Dr. Vargha-Khadem said.

In 1998, Dr. Monaco and a team of geneticists at the University of Oxford started to search for the variant gene, or for the lack of a gene, that presumably caused the London family's disorder. They identified as the likely source of the problem a region on Chromosome 7, the seventh of the 23 pairs of giant molecules in which each human cell's DNA is packaged. But the region contained about 100 genes.

Settling down to the long chore of examining each of these genes in turn, Dr. Monaco's team had a lucky break when Dr. Jane A. Hurst, the clinician who first examined the London family, came across an unrelated patient with what seemed to her the same pattern of impairment.

The new patient had a visibly odd version of Chromosome 7, enabling Dr. Monaco to identify a specific defective gene. The same gene turned out to be the source of the London family's disorder as well, although in the family's case it is sabotaged in a different way.

The newfound gene's product is a protein that binds to different sites along the DNA, signaling the cell to activate the nearby genes. Identifying the set of genes that is switched on by the protein could yield a deep insight into how a distinctive faculty of the human brain is constructed.

The new gene "gives us an entry point into language," Dr. Monaco said.

Discovery of the gene may give scientists a means to address the longstanding question of when language evolved. Although some experts claim to see evidence for language in hominid skulls millions of years old, others note the absence of another form of symbolic representation—art—until much more recently. If the two evolved together, language must be a fairly recent human acquisition.

Dr. Richard Klein, an archaeologist at Stanford University, argues that a specific genetic change produced the modern human brain some 50,000 years ago and that this neural change was probably the one that made language possible.

The newly discovered gene may enable geneticists to test that hypothesis by comparing the role of the human gene with the counterpart gene in chimpanzees. Chimpanzees can learn signs but lack the ability to string them together in sentences, a uniquely human ability.

Dr. Monaco said he had started a collaboration with Dr. Svante Paabo of Leipzig, Germany, who is studying the chimpanzee and other primate genomes. The two researchers will measure the rate of evolutionary change the newfound gene has undergone in the different branches of the primate tree, looking for any recent spurt in the human version of the gene.

Dr. Klein's idea that a language gene preceded human modernity reflects the theory first advanced in 1959 by the linguist Noam Chomsky, that language ability is innate, implying that there is a dedicated language organ embedded in the brain's circuitry. Others argued that language arises from an array of more general purpose cognitive abilities.

Asked about the new finding, Dr. J. Bruce Tomblin, a language researcher at the University of Iowa, said that several variant genes that seemed at first to affect only speech had turned out to cause other cognitive problems as well and that the same might prove true of the new gene.

"I am inclined to think there probably aren't such things" as language or speech genes, Dr. Tomblin said.

But Dr. Steven Pinker, a linguist at the Massachusetts Institute of Technology, said that he thought the new gene "shows there is some innate specialization of the brain for language" and that it provided "soft support, though not hard support," for Dr. Chomsky's thesis.

—NICHOLAS WADE

Language Gene Is Traced to Emergence of Humans

A STUDY OF THE GENOMES of people and chimpanzees has yielded a deep insight into the origin of language, one of the most distinctive human attributes and a critical step in human evolution.

The analysis indicates that language, on the evolutionary time scale, is a very recent development, having evolved only in the last 100,000 years or so.

The finding supports a novel theory advanced by Dr. Richard Klein, an archaeologist at Stanford University, who argues that the emergence of behaviorally modern humans about 50,000 years ago was set off by a major genetic change, most probably the acquisition of language.

The new study, by Dr. Svante Paabo and colleagues at the Max Planck Institute for Evolutionary Anthropology in Leipzig, Germany, is based on last year's discovery of the first human gene involved specifically in language.

The gene came to light through studies of a large London family, well known to linguists, 14 of whose 29 members are incapable of articulate speech but are otherwise mostly normal. A team of molecular biologists led by Dr. Anthony P. Monaco of the University of Oxford last year identified the gene that was causing the family's problems. Known as FOXP2, the gene is known to switch on other genes during the development of the brain, but its presumed role in setting up the neural circuitry of language is not understood.

Dr. Paabo's team has studied the evolutionary history of the FOXP2 gene by decoding the sequence of DNA letters in the versions of the gene possessed by mice, chimpanzees and other primates, and people.

In a report being published online today by the journal *Nature*, Dr. Paabo says the FOXP2 gene has remained largely unaltered during the evolution of mammals, but suddenly changed in humans after the hominid line had split off from the chimpanzee line of descent.

The changes in the human gene affect the structure of the protein it specifies at two sites, Dr. Paabo's team reports. One of them slightly alters the protein's shape; the other gives it a new role in the signaling circuitry of human cells.

The changes indicate that the gene has been under strong evolutionary pressure in humans. Also, the human form of the gene, with its two changes, seems to have become universal in the human population, suggesting that it conferred some overwhelming benefit.

Dr. Paabo contends that humans must already have possessed some rudimentary form of language before the FOXP2 gene gained its two mutations. By conferring the ability for rapid articulation, the improved gene may have swept through the population, providing the finishing touch to the acquisition of language.

"Maybe this gene provided the last perfection of language, making it totally modern," Dr. Paabo said.

The affected members of the London family in which the defective version of FOXP2 was discovered do possess a form of language. Their principal defect seems to lie in a lack of fine control over the muscles of the throat and mouth, needed for rapid speech. But in tests they find written answers as hard as verbal ones, suggesting that the defective gene causes conceptual problems as well as ones of muscular control.

The human genome is constantly accumulating DNA changes through random mutation, though they seldom affect the actual structure of genes. When a new gene sweeps through the population, the genome's background diversity at that point is much reduced for a time, since everyone possesses the same stretch of DNA that came with the new gene. By measuring this reduced diversity and other features of a must-have gene, Dr. Paabo has estimated the age of the human version of FOXP2 as being less than 120,000 years.

Dr. Paabo says this date fits with the theory advanced by Dr. Klein to account for the sudden appearance of novel behaviors 50,000 years ago, including art, ornamentation and long distance trade. Human remains from this period are physically indistinguishable from those of 100,000 years ago, leading Dr. Klein to propose that some genetically based cognitive change must have prompted the new behaviors. The only change of sufficient magnitude, in his view, is acquisition of language.

—Nicholas Wade